MASTERING FUSION 360 CAD-CAM AND CNC PROGRAMMING

The Ultimate Guide to CNC Programming for Routers and Turning Machines

By

Nishioka Yoshihiro

TABLE OF CONTENTS

FUSION 360 MENU OVERVIEW 5

OVERVIEW OF FUSION 360 NAVIGATION 18

FUSION 360 NAVIGATION OVERVIEW 19

THE POWER OF CONSTRAINTS 26

SKETCHING THE PART1 32

POST PROCESSING 39

PREPARING AND IMPORTING DXF FILES 42

EXPORT TO YOUR MACHINE 45

SKETCHING THE KEY 47

MODELING THE KEY 58

ROUGH MACHINING THE KEY 70

FINISH MACHINING THE KEY 84

PROFILE ROUTING AND POST PROCESSING 96

INTRODUCTION 103

TOOLS 107

TURNING 108

FACEING AND GROOVING 110

DRILLING AND REAMING AND BORING AND TAPPING 113

G00 115

G01 117

G02 AND G03 120

G04 123

G20 AND G21 124

G28 126

G40 AND G41 AND G42 128

G50 134

G71 137

G72 141

G73 AND G74 143

G75 AND G76 147

G81 152

G90 AND G92 155

G94 AND G96 159

G97 AND G98 AND G99 163

M00 TO M09 167

M11 TO M99 169

MASTERCAM AND COURSE INTRODUCTION 172

INTERFACE PART 1 SKETCH, MODIFICATION TOOLS AND TRANSFORM 177

INTERFACE PART 2 SKETCH, MODIFICATION TOOLS AND TRANSFORM 194

SIMPLE CAD DRAWINGS PART 1 203

SIMPLE CAD DRAWINGS PATT 2 211

3D DRAWING PART 1 216

3D DRAWING PART 2 222

3D DRAWING PART 3 228

LOFT ARC, SPLINE AND REVOLVE COMMAND 233

3D EXERCISE 1 240

3D EXERCISE 2 249

3D EXERCISE 3 252

SURFACES 272

MATCHINING 279

FUSION 360 MENU OVERVIEW

Before we begin modelling and constructing our initial basic part using Fusion 360, I would like to briefly review some of the features that we will be utilizing as router proprietors. Fusion is now equipped to accommodate machines of all sizes and shapes. We are capable of performing 3D milling and five-axis work. Tanning is feasible. Naturally, the majority of these are irrelevant to the operation of a 2D router. So these are the features that we would need to be aware of in order to use Fusion on our workstations. As we navigate the menu from left to right, we can examine the model. Here is where we conduct our preliminary sketches and establish the layout for our model during construction. Therefore, this is the feature that we would utilize frequently. We now understand that the sheet metal side will be required. Unless we intend to use it as a demonstration to promote our part, it is unlikely that we will want to convert it into a visually appealing graphic. We are not necessarily interested in animating sections for Iva. Therefore, the primary focus of our investigation is our model and manufacturing process.

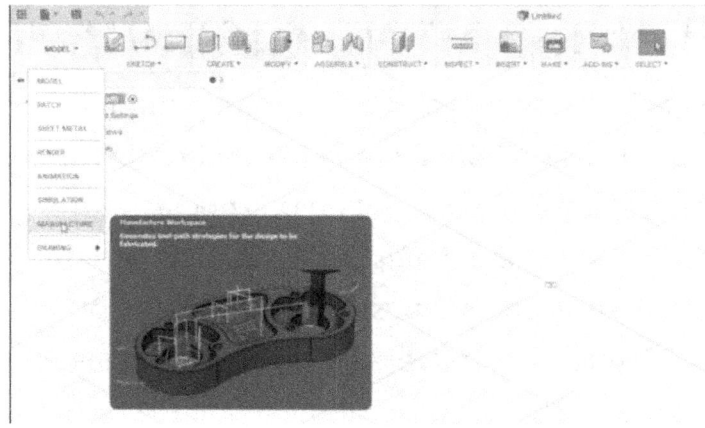

This is the two distinct aspects of fusion that we will be utilizing. We will be utilizing the model to construct and design our freedom model, and we will be utilizing manufacture to determine our two paths and cutters, as well as how we will remove the material that we have sketched during the model-making process. This involves changing the icons along the menu and changing the different workspaces. It is important to note that Fusion 360 frequently refreshes its appearance, so it may not be in the same location as it was when I first began building this course. However, the characteristics remain unchanged. Therefore, it should not be overly perplexing. A sketch menu is located in a model area and contains all of the various methods for sketching. The primary ones used in this context are the ones at the top. Therefore, the illustration comprises rectangles, circles, and arcs.

This is the method I typically employ to construct the 3D model and design prior to examining tooling plumes. It is always advisable to attempt to memorize the keyboard shortcuts, as they significantly expedite the use of Fusion when one hand is on the keyboard and the other is drawing on the screen. However, we can also add features such as Philip rats and the ability to trim lines. This is a valuable feature. I will quickly demonstrate this by pushing out to draw some lines on the screen. To see where these lines cross the crosshair, we may wish to trim them down. Therefore, in order to achieve this, we would either select the Tiki or proceed to a design and select trim.

Then, we would select the features that we wish to reduce the next intersection points by using the left mouse button. So, if we left-click on this hair with the trim option selected, it will trim the line down to meet the current one. We can repeat this process for the list line. The other features on the head are frequently used in mirror mode, which allows us to draw a line and offset it a specific amount to the left or right. Similarly, we can draw a feature and mirror it right across the centerline. Therefore, it provides us with a symmetrical component at the base, where we have provided the dimensions in schematic form. This is currently in frequent use. We can now apply dimensions to an arbitrary shape that we have outlined. After we have designed it and it appears to be in the correct location. Therefore, those dimensions will be generated by the machine always. Some individuals employ the shortcut key D to display our dimensions. Consequently, we can select two points, such as the two extremities of this line, and it will display a dimension. We can modify this dimension by entering into the box, which will result in a change to 50 millimeters. We can observe the results. I endorse this feature down to 50 millimeters. Consequently, we can create a preliminary design and subsequently add dimensions to ensure that it remains in shape, regardless of the amount of reduction.

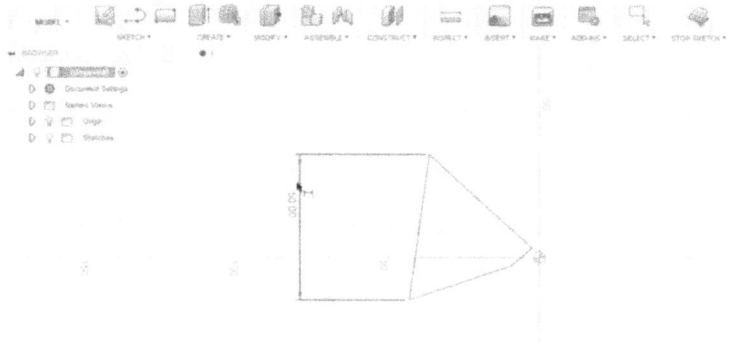

This dimension will remain at 50 millimeters in the future, unless we remove it. To rotate our pop, I will simply press the middle mouse button and the shift key. As you can see, it is still a flat design on a flat surface. As we progress from our sketch box, we are presented with the "Create" menu. This menu enables us to transform into a free-form drawing tool, allowing us to incorporate boxes, cylinders, squares, and other elements. However, I favor drawing in two dimensions and subsequently extruding it to create a three-dimensional object. It is the conventional method of conducting a CAT scan. I was instructed in this manner during my university years, which has resulted in our separation into independent components. We would require the extrude shortcut key to be used. To do so, click on the extrude icon. Currently, we have a small arrow that I am dragging upwards. This will transform our design into a 3D part. We can assign a dimension to this part, and we do not need to direct it—we can enter the dimension in the distance box. Therefore, in order to convert this 50 millimeters into a 3D element, I simply input 50 into the free D shape of my Freddy design. As we progress from the "create" option, we now have the option to "modify" the menu. This is the point at which we can begin to modify the arbitrarily generated shape we have just created in order to incorporate Phillip radius, which has two layers. We have the ability to incorporate shampoos, draw various features, and press and drag out of the 3D model. These concepts will be discussed as we progress through the course. This is a brief summary of the menu of elements that are relevant to our routing devices. To illustrate, I will apply a Phillips radius to the corner of this.

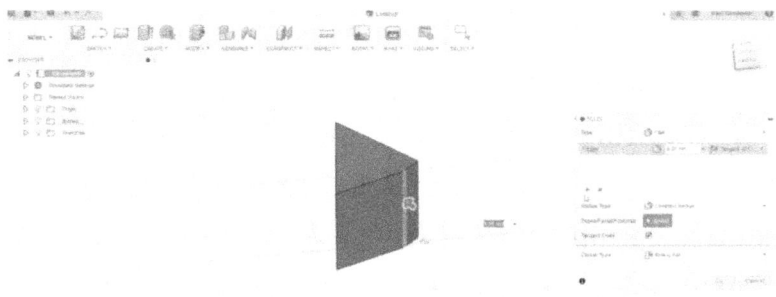

So I will simply click on fillets and then select a corner. I will select this one here, this edge here, and give it a four millimeter radius by checking the "Talk for an hour" checkbox and pressing the "Time" button. This will produce a four millimeter radius in a sharp corner between these two parts and some haziness. We can also incorporate the Shan phase and perform a variety of other tasks that are available in the sketch menu. However, we are currently working with Afridi objects and not a 2D design, so rapidly moving on to assembly is not a method that we rely on to gain a deeper understanding of the machines and their construction. This is primarily concerned with the creation of moving components that are compatible with one another. Consequently, we proceed to our inspection menu. This is one that necessitates examination due to the feature measure. Occasionally, it may be necessary to examine the measurements within our model. This allows us to click on the measure or utilize the ICC, which allows us to select two features on the model and receive precise information regarding their size. To ensure that the pie is compatible with the meeting elements for which it is being made, we can establish various tolerances. Each menu item will display additional options on the left-hand side, allowing us to determine the level of precision we desire, the units in which we wish to measure it, and the choice of snap points. Fusion consistently provides a plethora of alternatives; however, the menu panels located here may be left at their default settings. Most of the time, this simply grants us complete control over our component, which is an opportunity for experimentation. We have yet to observe the functions of each component. The majority of it will be addressed in this course. What I consider to be pertinent and significant are the aspects of our work. However, it is never detrimental to experiment with each feature to determine its impact on the subsequent item on our menu. Hair is derived from insects and is not particularly desirable; however, it is suitable for the purpose of capturing projects of cats. Other applications include the integration of FX files and solid vector graphics. These are the sections that I will discuss in various sections of the course. This is primarily to enable us to integrate a vector graphic into Fusion, convert it to a freely shaped object, and subsequently incorporate cutting components to facilitate its production on our machine. Nevertheless, this is a lesson that will be covered in the "Make" section of the course, which pertains to 3D printing. This is necessary in order to obtain a substantial shape that is devoid of any flaws, which the 3D printer can reprint. If you possess a 3D printer, you may print it; however, it is of limited utility for a browser. I will refrain from addressing that matter at this time. INS allows us to employ a variety of protocols and other resources. This section of fusion is quite advanced. I do not personally employ this feature when developing these types of programs; however, it may be necessary to investigate it in order to expedite your modeling process at a later time.

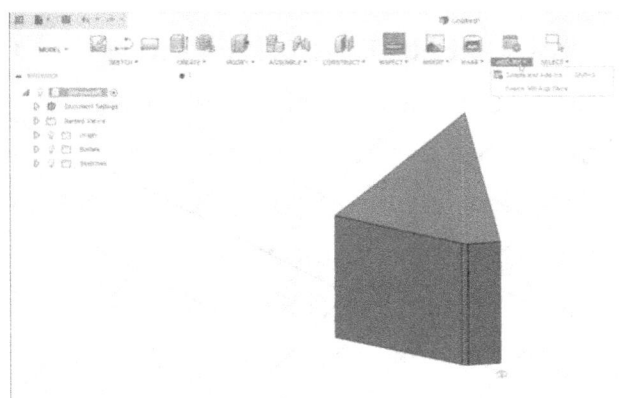

Please recommend add-on scripts that are available from Fusion to perform multiple duties with a single click. So, that is our model menu. We will now examine the manufacturing menu. We select the model and navigate to the manufacturing page. The manufacturing menu is displayed below. Let us begin by reviewing the setup process once more. Select "set up" and "new set up." This is the method by which we define the measure of our materials. Therefore, if we were to create these peculiar objects that I have haphazardly assembled, we would establish a bounding box. The material from which we will be cutting it. Additionally, this is implemented during our installation. Therefore, we would access the setup menu and select the operation stock. Therefore, we would typically consider milling for routing, and we have the ability to determine the extent of our work area and the orientation. This is advantageous if we consider the method by which we will secure this, whether it be through the use of fasteners, the voice, or any other system, in order to secure it to our machine. The next hub is either stock or stock. This is the point at which we establish the dimensions of the material from which we will be cutting our shape. We may add additional material to the sides as necessary. Therefore, we would employ stock offset and decide to add one millimeter to the top and sides of this material. However, the dimension of the box will determine how you desire to manipulate it.

Finally, we proceed to post-processing, where we can assign a program number and provide a program comment. Therefore, that is the configuration menu for the milling machine. This is the most pertinent option for us. This is the location from which we select our two components. To eliminate material. Next, we will proceed to the adaptive cleansing and the deep compartments. The two distinct techniques for adaptive clearing are illustrated in the diagram, which is used to eliminate material to create niches within our work space. It is highly efficient and rapid, as it maintains the same force on its hold at all times. The standard method of roughing material away from the inside of the pocket on a part is to face-to-face off the top of the material if the work piece is not flat. We may wish to employ a larger milling implement to clear the top of the material to achieve a smooth surface finish before starting to work on its activity. Contour is the routing method that we employ most frequently. This is the tool that we would employ to cut shapes from a flat piece of wood, whether internally or externally. If we were to cut a slot in our work piece, we would select a slot and the cutter would follow the central line of our design. Using the selection box that extends to the right, we can offset this to either side of the line. It may be necessary to employ circular once we have been chosen to utilize positions. This is beneficial for removing material from a round opening or for leaving a spigot in the event that we desire to cut around a spigot. Engraving is also a technique that we may employ frequently to incorporate text into our products, as illustrated in the diagram. Additionally, we have the option of a 2D humiliating effect, which involves shouting at the bottom of pockets to create a strong, attractive appearance. Finally, we have the 3D option of our menu. It is rarely employed in conjunction with routing. However, it may be necessary if complex parts with free profiles are included. Routing sequence routines are typically employed in a 2D environment, where they are used to cut shapes out of revel in the production of 3D shapes that are used as scenes. The milling machine is used for fusions on all C machines. Additionally, we can perform multi-axis drilling operations. We can also create programs for life. Currently, if your router is a plasma cutter or a water jet, we could implement a cutting sequence. This would enable us to create 2D profiles using lasers and plasma cutters. Once again, probing is more appropriate for milling or turning. This is the point at which we would employ an accurate instrument to measure the material or positions of our material or features on our model. For instance, we may have a hole that we desire to be dimensionally matched with another hole. In this case, we would create a probe and cycle to locate the center point of the first hole, and then use the measurements from that sensor point to manufacture a different feature. The actions feature is the final feature that we should address in this menu. This is the location where we can conduct simulations of our two components, generate the G code that our machine executes, and organize our post processor and setup documents based on the actions taken here. Therefore, this is the final step we take after the 3D model has been created and the two components are accurate. Next, we would simulate the process to ensure that it is accurate, allowing us to observe the machine's cutting action. Subsequently, we would post the simulation to convert it to a language that our machines can comprehend. Additional information regarding this topic will be provided in subsequent sections. Okay, that is a summary of the fundamental menu actions that will be employed throughout the course. In the subsequent lessons, we will delve deeper into each of these actions to ensure that you have a comprehensive understanding of them before we begin working with your machine and our 3D models.

OVERVIEW OF FUSION 360 NAVIGATION

Upon the initial launch of Fusion 360, there are a few tasks that must be completed to establish a working environment that is comparable to our C and C machine. This can occur at any point during the Muslim process if we fail to take action. However, I prefer to have everything organized from the outset. The manner in which I operate is more logical. Therefore, the initial consideration is our view cube. This is the method by which we rotate the components and select various orientations. Currently, the y-axis is not properly configured, as evidenced by the fact that it extends from the bottom to the top. Currently, the z-axis is located to the left of the spindle. In order to modify this Z, we must navigate to the preferences section by clicking on the menu item.

FUSION 360 NAVIGATION OVERVIEW

This refers to the opening of our preference box. We are currently in search of the default modeling orientation, as you can observe. It is significantly elevated. We only need to make a few adjustments to get started. Additionally, I would like to reverse the magnification orientation on this screen. Zooming out is more intuitive for me when I row the mouse toward myself. I am going to click on this in order to reverse the magnification, which will make it more in line with my perspective on where I can go. I am more accustomed to working in SolidWorks, and this is the way it operates. Therefore, it is more natural to me if I desire to modify the way everything functions in SolidWorks. I can utilize this dropdown box to modify it to that. Therefore, it is equivalent to an additional software component. Some U.S. However, I will maintain the setting of "fusion" for this 60-hour course, as it is a fusion-free option. So, once we are satisfied that we have configured everything in accordance with our default model orientation, which is the reverse in my case, we select "apply." As you can see from the view, the coordinates on the key have not been altered. However, we can adjust them by right-clicking on the front face. Set the current view to the top, as the axes are now running from the bottom to the top when the sets are rotated. In this instance, you have not updated the software; however, this is a software defect. This can be resolved by restarting the software at this time. Fusion 360 has been resumed

It is evident that the z-axis is now oriented accurately from the bottom to the top. It did previously; however, it is now displaying that it is present due to the fact that the x-axis runs across the front face, which is also accurate. This results in the y-axis extending from the front to the rear. Right now, there are numerous methods by which we can navigate our surroundings. Before we proceed, I will produce a model to provide a more comprehensive understanding of the operation of this system. Therefore, we introduce a model by selecting the height data panel, which opens. The following are some of the tasks I have completed in the past. Therefore, I will proceed with this one as a demonstration that is in near proximity for the time being. We have just selected the X in the most recent data box, and my model has been added. All right. The middle mouse scroll allows us to zoom in and out while holding down the middle button, which also changes our icon. This allows us to navigate our environment and track the model to view various parts along the x and y axes. By holding down the shift key and the middle mouse button, we can rotate up hearts by dragging them around the environment. Additionally, we can click on specific faces using our view. Please continue to monitor this location

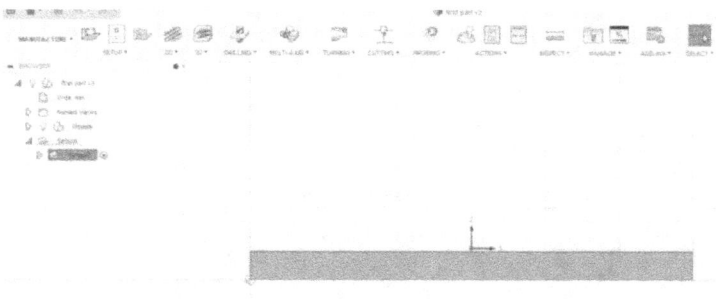

If we wish to view the front face, we can simply click on the face and it will bring us to the front face view. We can also rotate the cube by using the arrows provided, and we

can change the home position by clicking the hunky. We can select the faces on the sides and corners of the cube to obtain various views. To switch between these views, we can use the left mouse button. I personally prefer to use this method to view different parts of my part, as it provides a clear graphic each time I click on a different section of the view cube. These are the basic movements that I typically employ to rotate my parts and gain a comprehensive understanding of my work.

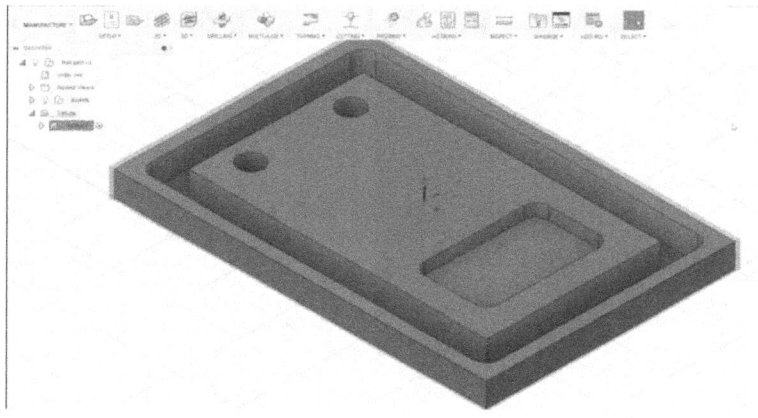

However, we also have this menu located at the bottom of the page. Now, this is also operated and controlled from a different perspective. Let us quickly examine this. In the Orbitz menu, we have the option to select "free" or "constrained." As demonstrated in the project, "constrained" is the latter. Consequently, we select it using our left mouse button rather than our middle mouse button. The pie's rotation is now much more natural, as the cursor has already been altered and we have tightened the constraints. For instance, it does not enter any peculiar positions when free orbit is implemented. It tends to occupy positions that we may not necessarily desire. Therefore, I frequently employ constraints to prevent myself from becoming unbalanced or difficult to manage. We can now select the visage we desire to view by examining the menu. So, for instance, if I wish to direct your attention to this visage, simply click on it, and it will be highlighted and brought to the forefront of our view. This is now applicable to any component. So, for instance, if we wish to examine the face in question, we can click on it and select "look at." This will bring us to the face. The "pan" button functions similarly to the "center mouse button" in that it allows us to move around the screen. We can also zoom in and out by holding down the left mouse button. The zoom window is also available. If we wish to examine a component in greater detail, we can use the zoom window to draw a frame around a cavity, for instance. This is Gates. We have the ability to immediately close in and assess its stability. This behavior is comparable to that of our software, such as Photoshop, where we would use a zoom window to zoom in. If we wish to apply a radius or roar, we get a much clearer view of the part we are looking at. Utilizing only the cursor button. I am able to extend out the middle mouse trigger once more, and the component now fits in a wider view on Zoom. I employ the F6 auxiliary key, which is quite sluggish. This is a valuable point to bear in mind, as it reduces the model's size to occupy the entire screen. I typically enjoy working in this manner when I am drafting or extracting the display settings for our parts. This now regulates the appearance of our various visual designs. Therefore, it is feasible to implement wireframe hair or tinting. I have a propensity to leave on shaded surfaces with discernible boundaries. It is the most straightforward to operate. However, if you require a more detailed view of the component, we can transition to wireframe, which provides a more comprehensive perspective. This will enable us to observe the situation more clearly. Therefore, I will return to the shaded mode with visual edges only and select Nelson, who is located below. We have the ability to modify our surroundings. For instance, if we desire a darker background, we can select "dark sky gray sky." This will alter the background, which may make the part more prominent depending on the configuration of your monitor. We typically leave this setting at "photo booth," which is the default setting. This provides a white background that can be used to display various grids and to attach to the grating for the majority of my sketching. You attach the grating, and I will add dimensions at a later time. However, we will address this matter when we begin programming the first section of Infusion 360. However, in this location, we have the ability to activate and deactivate the grates, as well as disable our cameras. That's great. Therefore, this is a viable alternative that we may need to employ at a later stage in the modeling process, as they alter our orientation in a manner similar to tapping on the faces of our viewer cube. Therefore, in order to view the front, I can click on two fronts, which will bring us to the front view. This is merely an additional method of utilizing our view cube. This concludes the lecture on the manipulation of views and the navigation of the Fusion 360. Additionally, we have addressed the process of configuring Fusion in preparation for the scene C machine by orienting the axes correctly.

THE POWER OF CONSTRAINTS

Before we proceed with the course, we will examine constraints. Constraints Infusion Free 16 is a highly efficient method for generating a design and roughing out our elements. Therefore, it significantly expedites the sketching process. Therefore, we will examine a few of the constraints and their operational principles. Therefore, we should proceed with this. I will commence with a top-down perspective. I select "maintain at the top" in my view. We can now create a clear sketch in this location. I will now press the "out" key on the keyboard. I will now proceed to draw a basic shape on the screen without considering the implications of my actions. This concludes our discussion. This is a shape that is entirely arbitrary. A shape can be approximated. We have a complete motion that is now over on a sketch palette here as it has cropped up. Presently, we have completed the design of our shape. We have the ability to access their limitations. Therefore, our constraints list is located here if we scroll down. So, let us examine the operation and nature of these. I will select the most pertinent ones to us at this time. There is a significant amount of information presented here that we would likely not utilize if we were to 3D model our elements for Scene Scene sequences. I am simply reviewing the ones that we would employ. Parallel is the initial concept that I wish to examine. Parallel is denoted by two parallel lines. As you may have inferred, this selects lines that are parallel. Therefore, if we select our constraints and subsequently select two distinct lines, such as this one, it is equivalent to the previous one. And this one over here has been parallelized. This constraint is applicable when we require a feature to align with another feature or two straight edges. This is due to the inclusion of an icon that indicates that the lines are parallel.

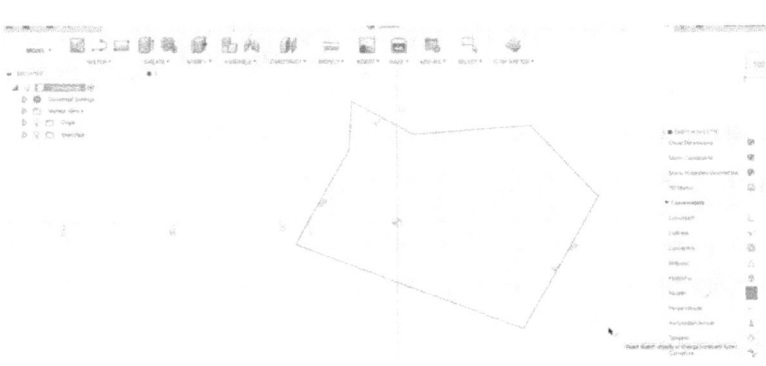

Another beneficial constraint that I will employ consistently is perpendicular. This allows us to choose two lines and align them at a right angle. Therefore, if we select the perpendicular feature and this feature, we can examine them at right angles in this corner. This is demonstrated by the perpendicular icon located here. It has also been demonstrated that this line and this line are perpendicular to one another. Therefore, they are not required to be joining lines. At each extremity of the components, we can choose two lines that are perpendicular. In addition, it continues to secure them altogether. Now, horizontal and vertical alignment is advantageous when we are endeavoring to align the sides with our datum position in order to achieve a smooth, straight edge weave along the Y axis or the x axis. In order to select this, we must click on the icon once more. For example, if we desire the horizontal or chill option, we can click on the line and it will be fixed in the horizontal position. Immediately, we can say that the entire section has been relocated. Additionally, it will encompass all other constraints that are in effect. Still, this is a correct angle. These two are also perpendicular, and they remain parallel to one another. Additionally, we have equals, which are the extent of each side or line of feature. Therefore, in order to ensure that these two lines are of equal length, we would select "equals" and subsequently pick our two lines. Additionally, it has resulted in him being precisely the same length, which illustrates the concept of concentric and tangent lines. In order to generate a new illustration, a neutral airflow is required. I will navigate to a file and a new design, and then click on the top of the view keep to secure the perspective from the top down. I will be drawing circles this time. So, we could observe the keyboard for this purpose. Our data position will serve as the central circle. He is going to click on that, hold down the mouse button, and bring out our circle. I am aware that I will be proceeding to another individual adjacent to me, who scrolls down a sketch palette to our constraints and selects "concentric." This time. By clicking on each circle, we can now view them in conjunction and designate them as concentric. This is an enormous time-saving feature when it comes to creating lifting eyes, irons, or any other application that necessitates a cavity with a corresponding radius around the perimeter. Now, in order to achieve a harmonious combination from this radius, we will implement our tangent option. In the sketch palette, the constraint is that we can link a line to create a seamless, curving curve that terminates at a precise tangent to the radius.

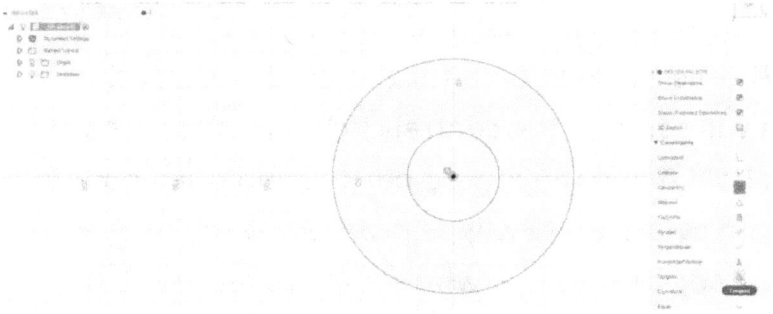

Therefore, in order to illustrate this, we will press the "out" key on our keyboard to elevate our line. We will then draw a series of lines that emanate from our radius. After the line is finished, we press the escape key to conclude the drawing, and we repeat the process on the other side. Therefore, our four lines will create a line in his escape to cease drawing immediately under our palette's constraints. We possess our tangents; therefore, in order to facilitate this, we must select the attention icon. We select the radius and the line that we wish to be tangent to it. It appears to be a uniform curve at 10, and it has also provided an icon to indicate this. It is currently locked in place; however, we can relocate any feature whenever we employ constraints; however, the constraint will remain in place until it is removed. I will repeat the process this time. I will select on the line and our radius, and it will provide us with a lovely tangent. Therefore, if we were to machine this, we would not have a step, as the line defines the radius of the part. Therefore, if we were to complete this feature, we could remove the superfluous lines that are unnecessary. To do so, we would press the "t" key on the keyboard and then select the area we wish to remove. So it is similar to this radius, and it has left us with a lovely tangent that will curve into a straight line around it. Therefore, this would be a machinist that does not require any stages and would be aesthetically pleasing. Once we have completed the sketching and have exhausted the constraints, we can proceed to Las, where we can employ trim in conjunction with these constraints to create a visually appealing feature. From there, we can simply click on the "stop sketch" button located at the bottom of the sketch palette and begin the extrusion process to convert the sketch into a 3D model. So that's a concise summary of the primary constraints we encounter when employing Fusion 360 as a C and C machinist.

SKETCHING THE PART 1

In this course, we will examine the process of constructing a basic pot infusion 360. I would like to provide an overview of the software's functionality and how we can utilize it to create a container for our ounces. This is Fusion, and it has recently been opened. I prefer to work from a top-down perspective. Therefore, by selecting the center of the view and maintaining it, we are presented with a clear top-down perspective that allows us to observe our work directly from the machine's table. I will simply maintain the middle mouse trigger and manipulate the screen.

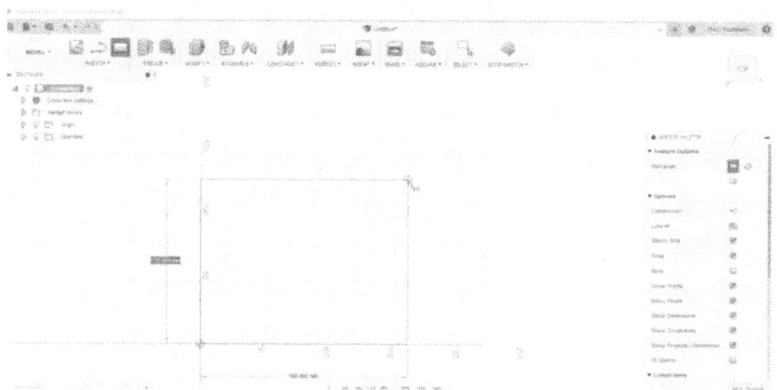

Therefore, the location of our starting point is lower down. I would like to commence by sketching the working area or the material that we will be cutting from a 200 millimeter square piece of plywood that is located in a large cavity. This is the reason we ascend to our sketchbooks, where the openings are two-point rectangles. We have the option to either hoist our coolies up or click on this link. So, proceed to click on the two-point rectangle. We will begin by establishing our reference points in the lower left-hand corner. I will begin to draw our material from this location. To do so, simply left-click and bring up the hub box. Now, while holding down the middle mouse button, we can navigate around the area to gain a more comprehensive understanding of our working environment. So scrolled out and zoom out using our middle mouse will say if we come up to here I've actually got a snap to grid locked in at the moment so snap in two different ports parts of the great so I'm going to click this here and that's sketched out the working area I can just push escape Kate come out of our sketch mode now holding down the shift key and middle mouse button we can rotate our part at the moment it's still a two dimensional sketch so that's not a problem that we need to make this into a thick Police piece of wood and we did up by coming up here to create we want to extrude the shortcut key for this is E or we can just click on this icon here to extrude our material into a more solid looking part so we're gonna want this to do this to ten millimeters so we can click this blue arrow in the middle here and bring up material like this or we can type in this box the size the material US type in ten point zero MM and click return. We now have a 3D representation of the material that we will be cutting. Fusion operates by requiring us to sketch each individual part one at a time. Consequently, we would render a feature, make it free day, extrude it, push pull cutaway to the material we do not wish to use, and then proceed to sketch the next feature. We would not do this simultaneously, as we would lose the features we have already sketched the first time. We would begin with the first feature, which is Street. Now that we have our free material, I will rotate it using the shift and middle mouse button. We can begin sketching on the surface of the material we wish to remove. At this point, we are only examining a basic part. Therefore, I will remove the material from the middle and create a pocket to do so.

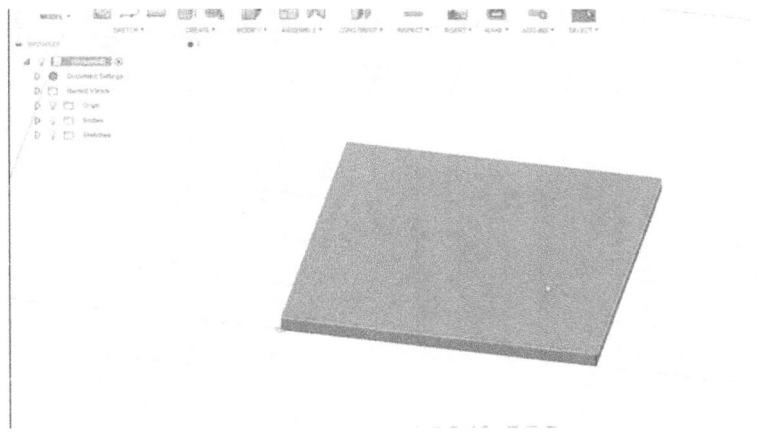

So we're left with this 200 millimeter piece of material here we have some shapes removed from the middle so let's click on the top five you keep to snap it to the top down position I'm just going to randomly sketch a square no particular four and mind four dimensions at this stage we're just looking at the process that fusion expects us to use when we're modelling it starts off our sketch around here and by pushing out to bring up Lowenthal or we can click on this icon here for the same thing sufficient hours perhaps so I'm going to click on this part here to start I'll join now I still have a snap to get selected normally if we're working to tight dimensions we wouldn't use this I'm just using snap to grid at the moment to give us a quick basic shape we can sketch out just for demonstration purposes we will move onto more complex shapes and patterns as the lessons progressed so as you can see I'm snap into a grid here some get snapped to this point.

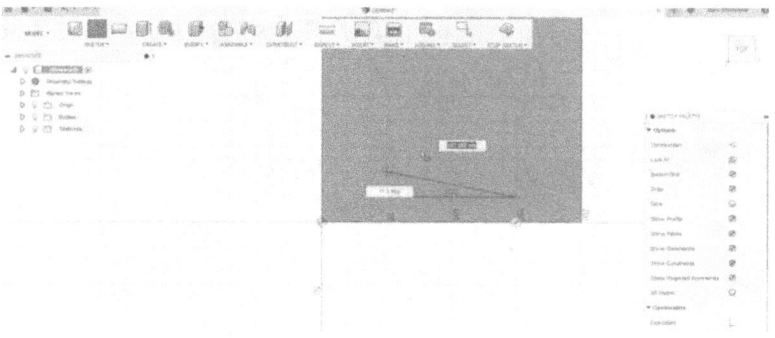

We are now able to autonomously draw a straight line. It provides continuous information regarding the angle and distance, which is extremely beneficial when designing components with this software. To create a radius or curve, place the cursor over the initial point of a line, hold the left mouse button, and maintain the button press. As we distance ourselves, it is evident that a radius has been drawn, and it will align with the individual we wish to observe if we move in squares in one direction for approximately two squares in the y direction. We are aware that the mouse button will be released, causing the object to lock into place. We can proceed to draw the next part by clicking here and moving the cursor 50 millimeters to create another straight line. By holding down the middle mouse button, we can move the parts around, which can add another radius. The object will then move with a small curve, which is too beautiful to bring up a feed into the next line. This results in the formation of an additional radius. Therefore, we accomplish this by selecting Holden on the final evening, which will expand our radius to the desired location. I will release the cursor at that location, and it will perform a smooth, blended curve to our position, rather than a direct line, followed by an additional radius. I will once more select a 20 millimeter radius and release the cursor button to secure it in position. Lastly, I will complete our shape with a 20 millimeter radius. Therefore, I will proceed to squares that way, two squares that way, and complete a straight line at this time, as I am confident that this will result in a pleasing radius. Therefore, it is well-integrated into the linear lines. We are currently in the line mode. By pressing the escape key, we can exit that situation. Sara is informing us of the tangential parts by the right angle portions, which are parallel to each other, as you can see from the constraints. Consequently, we have our desired geometry. As we discovered while reviewing our constraints palettes, these are automatically identified; consequently, our preliminary design has been superimposed on the material. We begin by removing this component by selecting the "extrude" option and pressing the "e" key on the keyboard or clicking the icon. From there, we can select the elements from which we wish to remove material. Therefore, we would like to eliminate all compartments in this area by employing a solid click. In the same manner as when we initially arranged our material on the table, we can manually advance or retract it to separate this section. Therefore, I intend to drill a depth of ten millimeters, as our material is 10 millimeters thick. Therefore, we have a greater clearance at the bottom to carve the shape. So I will enter minus 10 points in the dimension field and press the "10" button, which will result in the cutouts being displayed in our profile. The subsequent section of this course examines the cam sides, where we manufacture and determine our tooling components. This concludes the straightforward sketch lesson. In the subsequent lesson, we will examine the camera aspect of the Catch CAM program.

POST PROCESSING

We have now completed the model and have generated two passes. The subsequent step is to export the model to a language that Valter can comprehend. Let us proceed with this. I am currently able to witness the poster procedure, which includes the information that we must be aware of. Initially, it is necessary to determine the machine that will be used. We maintain a list of devices that generate two lines of code. It may be a code for please ones at the top of the page or a height in meters located at the bottom. Therefore, the types of languages that it employs are distinct. Additionally, we are experiencing great difficulties. However, it is probable that the G code is being utilized by the relative who is otherwise located in Cincinnati. Therefore, it is necessary to choose the item that is closest to your router, which is a lengthy hair.

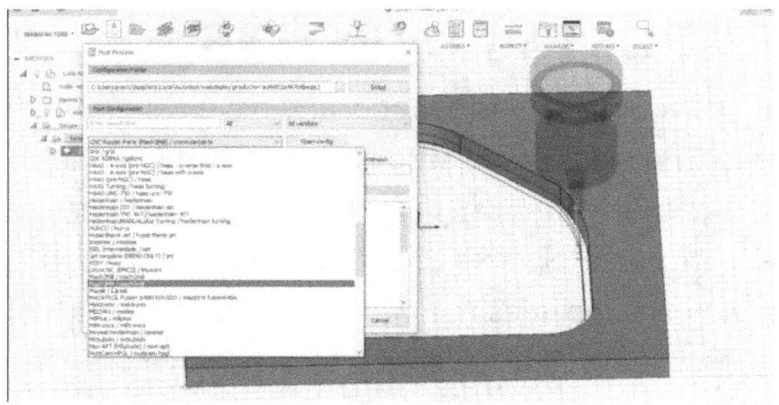

I am now going to opt for Mac free, as Mac FRE is a prevalent option. Therefore, I will use this post solely for demonstration purposes. Therefore, we will proceed to select the Mac that is currently unoccupied. All g codes are essentially similar; however, setting devices employ distinct customized codes and anticipate distinct languages. Therefore, if there are only a few machines that have been used, I encourage you to experiment with a variety of them. It is likely that the machine manufacturer has a post processor that you can use to import into an open Fusion 360 to convert your parts into code. The output folder name is the location to which our file will be outputted for our router. We can then provide a program name or number and comments. For the time being, we will leave the value at one thousand and one. Additionally, we can select our units, either in inches or

MMS. Currently, the majority of machines can operate in both modes, and a single line of code can inform the machine which mode to operate in. Consequently, I will opt for MMS at this time. Currently, when we are prepared to post, we simply click on the "post" button. The system will prompt us to save their phone numbers, which some individuals refer to as "1001." We then click on the "safe" button. Fusion employs brackets to display our programs. This piece of software has been unlocked. Brackets are frequently employed by programmers to compose programs. It is possible to observe the appearance of our program if I expand that. This is a program that can be accessed by your Cincy router. It is not exceedingly lengthy. The program is relatively brief. So, in a separate section of the course, we will dissect this program and provide a line-by-line explanation. However, at this juncture, we can simply transmit this file to the Frieda router, which will subsequently sever our path. This is the post-processing phase of Fusion 360, which is responsible for the production of our functional G code file from our freedom module and its subsequent transmission.

PREPARING AND IMPORTING DXF FILES

In a sequence of lessons, we will examine the process of generating components solely from the Excel file. This is a vector image file that can be imported into Fusion 360. Next, we will generate a tool pass to enable the production of the components that we generated using our vector image. Therefore, Illustrator will be the software of choice for this task. The method I employed to create this component was to display a photograph that I had previously traced around the scratch plate of the instrument. This allowed us to generate two plumes. There are numerous outlet software programs available that may be used to generate vector images. Therefore, in Adobe Illustrator, it is possible to export a smaller file to the x F format, which is the standard file for AutoCAD. AutoCAD actually created this file type as a fusion. Therefore, it is highly compatible with Fusion, rendering it an advantageous option for our utilization. If we select "export" and "export as," we can then select the "effects" file from the drop-down menu. What we are seeking is the vendor's scratch plate, which we will export under its name. Therefore, that is the result of the f x file. We must now import Fusion into the fusion load loaded up. It is imperative that we prepare the environment for the import of our file. The initial step I will take is to select on the top of our view keep to display a two-dimensional top-down view and enter the location at the SFO. We received two items and subsequently inserted them into the menu box. Therefore, we must access this document that requests that we perform tasks for free. It requires us to determine which plane we will be importing this sketch into. Therefore, I will select this operational plain. Therefore, it occupies the same plane as our machine table. We must now extract the Excel file. To do so, click on the icon and select the file that was created in Adobe Illustrator just a few minutes ago. The file contains guitar parts. It is now necessary to place it in a location that is suitable for use in the workplace. Therefore, I would prefer that the data be located in the lower right-hand corner. Then, we can position our words so that the data position is 0 0 in the bottom corner, and all dimensions will extend from that point. This is our machine data in position, and we scan and drag out parts. This allows us to import the width and perform a motion measurement of the part. This is the position of our data along these axes, and we can zero it on the machine.

Initially, Sally will ensure that X and Y are zero, and subsequently, the remaining components will be manufactured. Components from those positions. So once that is completed, we can simply select "okay." Additionally, the appearances are in alignment. Therefore, we have requested a scratch plate for a machine table; however, this is merely a preliminary design at this time. Therefore, it is necessary to extrude it in order to transform it into a three-dimensional form. I will rotate the working environment by pressing the middle mouse button and the shift key, and we will select this icon. The extrude icon is accessible by clicking on the component and utilizing the arrow to extrude in any desired direction, as the scratch plate is extremely thin. I am going to increase the size of the scratch plate by six millimeters. To do this, we will enter six point zero in the box and press the return key. This will add a small amount of volume to the scratch plate. So immediately, we have transitioned from a low-fidelity drawing to a real 3D part that we can begin investigating for the purpose of constructing tool parts. Therefore, in the subsequent instruction, we will develop a program that generates this effect on our relatives.

EXPORT TO YOUR MACHINE

Therefore, it is time to convert this into a g code program and export it in a language that our relatives can understand. In order to accomplish this, we select the post processor icon, which is represented by the G1 and G2 symbols on the interface. Currently, there are a few items that require selection. The primary consideration is the machine to which we can output. Therefore, we retrieve this list and proceed to locate your machine. If your machine is not present, please contact the machine supplier, as they are likely to have a post-processing file for Fusion 360. It is an exceedingly prevalent software application. We encountered proprietors, and post-processing is typically available for your specific control.

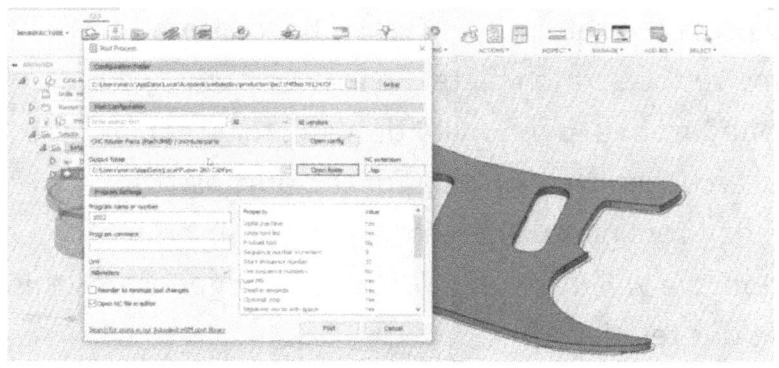

Now I'm gonna stick with using Mac free because Mac free is one of the most common used ones so low I can't cover every single machine Mac free is a good start one to use this as a demonstration you select since he rounds a parts Mac free scene C round two parts that's the post-processing name that's available to us now I'm going to give it a program number of 1000 and free changing units to MMS now we also need to select where the file will be safe to on APC so I'm just going to click on this icon here and save it to my desktop and now we're ready to post this program and save like sufficient to save that and it's opening up the brackets right now which is the software fusion has built in to view these case files have a quick look through a file which make sure everything's right we have our safety line at the top here and then it goes into the 2D console where it gives us all our positions in g code g code file looks complete. I will now dismiss that window, and we can proceed to upload our files directly to a machine to observe their functionality. Therefore, this is the procedure by which we generate the Excel file, upload it to Fusion 360, convert it to a free object, and subsequently generate our two paths. We then employ the post processor to transform the object into a language that our routers can comprehend.

SKETCHING THE KEY

It is part of a sequence of courses. We will examine modeling that is slightly more intricate, incorporating a variety of topics that we have previously discussed and some that we have not yet addressed. Occasionally, individuals present keys as a gesture of goodwill during their 21st or 18th anniversaries. Therefore, we will simulate one of those keys and convert it into a program that can be cut on our CMC Walter. Okay, Fusion 360 has been initiated, and this is the beginning of our illustration. I will commence by selecting the uppermost view on our view cube. Therefore, we are afforded a pleasant perspective from above. I am in search of a module that is approximately 200 millimeters in height. By a length of 400 millimeters. Initially, we will utilize our data to position Hampton and maintain it in the lower left corner to ensure consistency across our programs. Subsequently, we will draw a circle. Therefore, it is imperative that we generate a design. This commences with the creation of two-dimensional sketches. We must determine which aircraft we will be using to illustrate. Click on our plane here. Moreover, it is providing us with our design palettes, which include our constraints and various options. We can temporarily close this out. My initial action will be to join the circle. The primary affirmative key. Therefore, if we select "Create," we will be presented with a circle and a variety of alternatives for the creation of a circle.

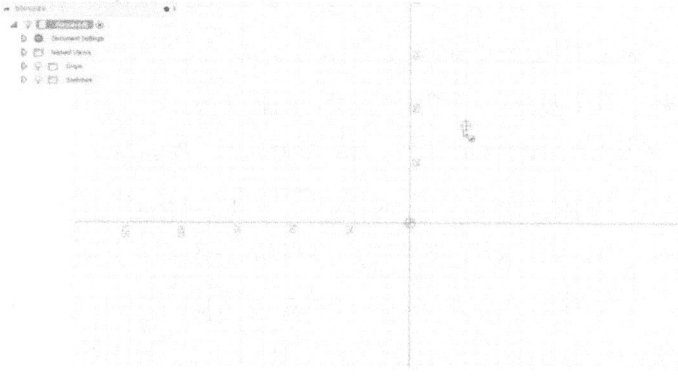

I will now utilize the sensor time to create a 200 millimeter diameter circle. Alternatively, we can press the C key to create a shortcut or simply click on the object and select it. To do so, use the middle mouse button to navigate around and determine the desired location. I will commence at this point. Locate the "sense point" of our circle and extend it to a distance of 200 millimeters. We can simply type two hundred and press the return key to secure it in position for someone we have now circled around. It is evident that I have accurately calculated the sensation of pointlessness. I believe that the center point is at least very precise, as it intersects the R2 datum lines. I would like to temporarily relocate that item. We have a material that we can cut if we use a datum as the corner of our material. To move it, simply right-click on the center of the circle and select "move." Duplicate. This provides us with the ability to reposition it by utilizing the arrows. I will simply reposition it and secure it slightly away from the extremities of our datum position. It may be beneficial to activate Snaps, which I have already done. This simplifies the sketching process, particularly since we are not focusing on geometry or dimensions in this instance. We will be focusing on the aesthetic appeal of the design rather than the precise dimensions.

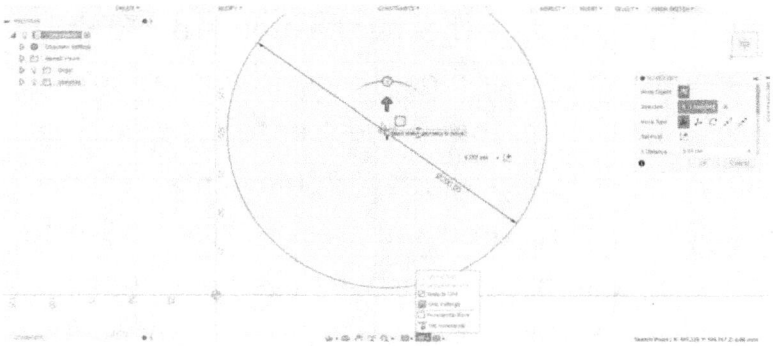

Therefore, it is comparable to that. That is acceptable. "I mean, simply click "agree" to secure that." He is the middle mouse key that is used to pan out and navigate around. I am going to create a line from the sunspots that circle, which we can use for construction purposes. So, a shortcut key that could draw a low income away from the area is to click on a line tool or press out. At approximately three hundred fifty millimeters, it is feasible. It is possible that we can enhance the appearance of this 400 400. After entering 400, we have a 10 in the field, resulting in a center line. We will no longer be pruning along this line. We will be employing it solely for construction purposes. This is a construction line, and we can offset it to construct our design by simply offsetting the objects. A simple shorthand technique involves pressing the key, which will display a list of sketch shortcuts. We may commence the process of inputting any Fusion 360 option into this field. For instance, if I wish to offset and commence typing the term "offset started." It has appeared at the top and displays the icon. The icon offset is also present. I will select "offset" and then click on the center construction line that we have created. I can then drag it in either direction to offset it. Based on this, I would estimate that it is approximately 45 millimeters. Therefore, we are considering a negative 45. Let us ensure that this is precise and eliminate the decimal digits. I will simply enter minus 45 here and here at 10, and I will repeat the process in the opposite direction with our line here. Therefore, the icon is offset to the right, and I will move it down by forty-five. Therefore, it is identical. We could now remove the construction line from this location; however, I intend to leave it in place for a brief period. Therefore, we have a reference for the centerline of the part. If these dimensions are causing confusion for the viewer and obstructing our progress, they are not necessary at this time. Thus, to eliminate these dimensions, we can simply click on them with the left mouse button and press the delete key, which will eliminate the dimensional lines displayed. This provides us with a more comprehensive understanding of the object we are modeling, which is a benefit of Fusion 360 or any other CAD/CAM software that integrates the sketching or 3D modeling stages. We have the ability to make modifications at any time by examining this. I would like to relocate these lines slightly closer to the central line. That is not an issue. We are capable of accomplishing that. Here, you may select the offset icon. The offset icon allows us to move and drag this line to any location. Therefore, I will reduce the measurement to 40 millimeters. We can only hope that it is included in this, as it would result in a 40-hit return and a five-millimeter reduction. Additionally, we implemented an absurd conclusion. Perform a double-click on that. Modify this.

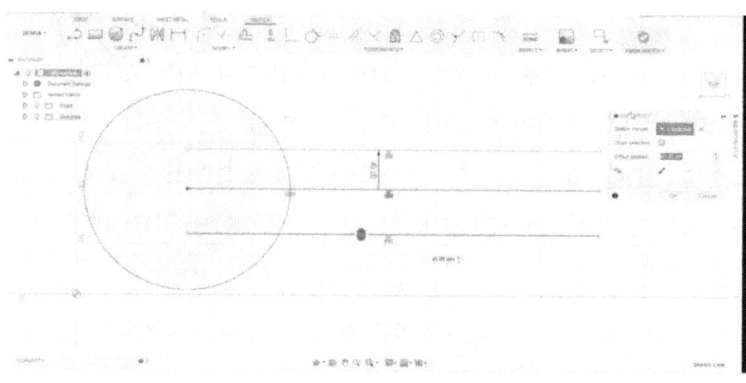

We have the ability to modify it in the edit box, as well as some square demonstrators that are discussing 40 millimeters and head return. Additionally, we can eradicate these dimension lines by simply flexing them and pressing the delete key. Now, I would like to introduce you to one of my favorite tools, Infusion 360. This tool is particularly useful when we are not relying heavily on dimensions and simply want to achieve a specific aesthetic. We can utilize the fixed point spline icon that is displayed above. This allows us to create a beautiful curve shape. We must now consider the radius of our curves when designing them, as the cutter must fit within the radius and be smaller than it. To enable it to be machined. As we are designed in segments, it is necessary to consider the method by which we will reduce this. So, if we select on this icon up here to draw our fixed point spline and I want to draw a small feature here that some keys have, we must be very careful to ensure that we do not make the radius too small. So, let's zoom in a bit and move the center of the screen as we draw this section. There is no such thing as freehand. I do not intend to replicate any image; rather, I am aware that children frequently possess this component of some design. If we are dissatisfied with it, we may modify it. So, when we select a point on the radius and click in various locations, the line automatically generates a radius. We can then move the points around and select a node by clicking on the screen. The node appears to be in place, and we can continue to draw around the curve. I am simply drawing something along the lines that this individual has drawn.

Perhaps a lovely radius should be incorporated into the area. There we go. Then, when we are finished, we simply strike a tanky, which appears to be completely in position. Therefore, if we expand the image to observe the appearance of that. Certainly, that is the type of feature I was seeking, and we can always make the contours merge more smoothly. Once the 3D modeling stage is complete, we can eliminate this bottom line. I propose that we execute this action. To trim a line, simply press the "t" key on the keypad. The selected line will be removed, and we can proceed to trim the unnecessary line. Now that we have the key outline at the top, we would like to maintain the consistency of this feature due to the fact that it is not directly at the key. I will create a tidy radius within this area, and in order to do so, we will determine the direction of the radius. OK, so we have reached the arc, as Fusion refers to it, and we have a few options. We obtained free points, arc center points, or tangents to the arc. I am going to stick with the free points arc at the stage, so it will be similar to the point he had in there, and as we draw it down, Joseph will have a lovely free point. Therefore, the initial and final coordinates were inserted, and the ark was subsequently manipulated using the mouse to determine our preferences. Therefore, the appearance of that location is quite appealing to me. Clicking on this mouse icon secures it in position. Fusions now convert this into a solid component. This is no longer an issue, as we continue to augment it. We can delete this line at a later time, and I will eliminate the connection between the two and create a single key that is connected. Next, we will construct the TIF part of the bottom of the key using the same fixed points applying tool as before. We will use this line as a guideline and can go over it. Our goal is for the entire central line of the bottom motif to be a roundabout, matching the top part. We do not want the key to loop off center or to one side. Therefore, this is more of an artistic rivalry in the realm of design at this juncture. To begin, we select the end points of our radius and the line that represents our fixed points. I refrain from bringing the chief down at this angle, rather, about there. As we descend to Joe's, we should ensure that the radius is well-integrated into the bottom line. And we must keep this in mind as we proceed. The radius is too narrow, and the cone is too large, necessitating the use of a cutter. We do not want to enter the machine with two small diameter cuts, as the finishes radius is off. To ensure that it is easy to machine, we can slightly increase the size of one of the arcs by clicking on the points as I am drawing the TIF. This will allow us to insert a cutter and lower the other one slightly. Additionally, we must take into account the inside radius on this side. The reason for this will become apparent in a moment, as we will offset the key as soon as it descends to the bottom, and I will endeavor to seamlessly integrate it into the existing design.

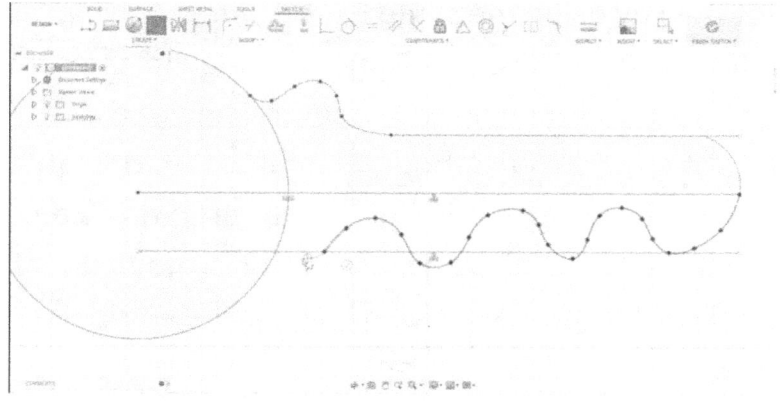

So I proceed from that point onward, where it merges in, and click on it. I then press the return key. Once we have completed the task and eliminated all of the lines and instructions, we can simply select anywhere on the screen to eliminate them. As you can observe, the lines at this location are interactive, and it is uncertain which objects are substantial and which are not. Therefore, we can proceed to remove these lines. Currently, we can accomplish this by utilizing the trim tool or by pressing the Tiki button, which will activate the trim icon located above. In order to ensure that we can only see Aki's teeth, we can simply trim away these lines and remove our construction lines. Now, there is a construction line that is visible on the screen as it scrolls. Therefore, we can also eliminate that. You have received a warning; however, it is merely informing you that we have eliminated all of our construction lines. Therefore, that is the QI shape that has been established thus far. We must now begin the process of converting this into a three-dimensional object. Therefore, before we begin transforming this into awe-inspiring objects, I will simply remove the remaining lines using the trim to 30 tool. I will select the lines that need to be trimmed before I click on them. We have now completed the outline of our key. While it would be preferable to have the composite a bit smoother, this can be accomplished during the 3D modeling stage. There is one additional feature that I would like to relocate, which is this curve. We do not require this sound, so click on it to delete it. All right. We can now proceed to the subsequent instruction, in which we will begin to transform this into a three-dimensional shape.

MODELING THE KEY

Flares are indispensable. The sketch has been completed, sir. We will click on the "finished sketch" button to proceed. We can now begin to examine the model in its entirety in 3D. Therefore, in order to rotate our planar design, I will press the shift key and the middle mouse button. I am now going to extrude that. I will opt for a thickness of 15 millimeters. Therefore, if we reach the extrude icon or press the key for our keyboard shortcut, this will transform into a solid shape. We can then follow the arrow upward to transform it into a 3D shape. I am now going to aim for 15 millimeters, or perhaps twenty-seven millimeters. Let's call it 20 millimeters. I will convert this into a 20 millimeter thick component by entering 20 and pressing the return key. The outline profile of our key is now available. Now, we discussed the possibility of establishing this radius here; however, Smith has to intervene and complete the task. Therefore, the method by which we can accomplish this is to select the feature that we wish to modify, which is the imaginary line here. Subsequently, we can transform it into a flat right Sangre by pressing the keyboard and springing up our short khaki. I do not wish to have Philip right. I enter "fill it" and subsequently, we have a sturdy Philip. In an illustration, we can also implement it as a 2D shape. We desire the object to observe this if we are employing a three-dimensional shape. Therefore, you select the Phillips icon located here. We can now manipulate the shape of this by dragging it out and reintroducing it. We are solely interested in residing in this location, so I would estimate that these 26 millimeters and a round that is off by 25 millimeters were examined in place with the retained key. The result was a radius, and as we emerge, the curve is much more aesthetically pleasing. So that appears to be a great deal simpler and more aesthetically pleasing. It is a machine, and we can verify that none of these radius signs are too small for our cut. We are acquiring this one, which may be slightly too small, but we should be able to handle it for the time being. We can always revisit and revise that at a later time if necessary. All right.

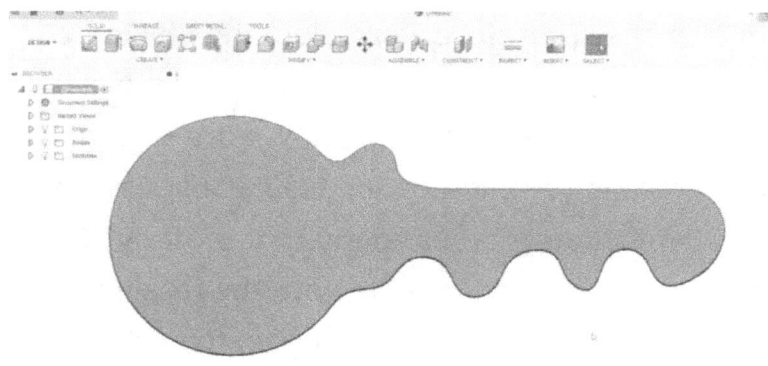

Basic outline of our key is comprised of slats. We will now focus on the internal features and effects that I am attempting to accomplish. My objective is to create a step that extends around the exterior of a recessed pocket that is approximately 10 millimeters deep within our components. Therefore, it is a straightforward method to accomplish this. To obtain a more detailed 3D view of our components, maintain the middle mouse key while holding the shift key. I will offset this external profile to provide the geometry we require today. This will initiate our sketch show cut, and we will begin with offsets. We will then select offsets at the top. It is now sufficient to select the exterior profile, which will provide us with a slide key to adjust the line to our desired position. I am currently contemplating a range of four to five millimeters. I am still cognizant of the radius at this location when it is sufficiently large to accommodate a cutter. Consequently, I will increase it to four millimeters.

We now know the direction in which we need to move X, as it is displaying a minus sign in the Offset position. Therefore, by entering minus 4 and pressing the return key, the provided line within the specified radius will be satisfactory. Acquire a cutter at this time. That is precisely the type of item I was seeking. Therefore, we can now extrude the material and transform it into a desired shape, which we can then remove. That has been disclosed. The sketch has been completed, and I will extrude it once more to enable us to activate the shortcut key e or this icon. I will use the shortcut key this time and select any location on the face that we wish to extract. This will be done before the little arrow appears, allowing us to drag it up and transform it into a three-dimensional shape. Now, when it decreases by 10 millimeters. Therefore, we converse in the distance box (minus 10) and strike 10, which results in the cut. Now that we have a smooth step encircling the key, we can route it out with our tools while we are machining to create a three-dimensional shape. I am considering the machining process in advance. We will expedite the internal work and carve the shape last. All right. Now, I would like to incorporate some features in the center of this area that indicate that it is an individual's 21st birthday. Therefore, we will assign the value of 21 to this section of the key. I propose that we examine the matter. In the event that we are uncertain about the desired feature or do not wish to display the location for its selection, we can incorporate it into the menus. We are once again encouraged to utilize our design conveniences, and I am under the impression that this is a positive development. Perhaps we should refer to this as "text." Therefore, it is not entirely contained within the text, and it is evident that there is a feature in fishing 360. So I send a text message indicating that we are going to click on that and select our working aircraft, which is the one listed here. Therefore, we will choose that option. Therefore, that is the surface that we wish to address. Currently, we can select a location in the center of the screen and enter our text on 21. Presently, it is inverted. So it is internally oriented, which allows for a 180-degree rotation. In fact, we prefer it this way, as it will be two hundred and seventeen. It is somewhat diminutive; therefore, we may modify our expectations regarding the textbooks. We may experiment with hundreds of them to determine their appearance. It is possible that the current size is slightly inadequate; however, we can locate the origin point and relocate it to increase the size by approximately one hundred fifty. Observe the appearance of that. In fact, I would say that it is ideal, as it is getting closer. A substantial amount.

Therefore, we can consume that and relocate to the desired location. I will adhere to that. We are now able to position the tool between the perimeter of the bars and the one in question by examining the machine and process. It is possible that it is slightly larger than expected, as we intend to extract the entire fruit. Therefore, we should reduce the size to approximately one hundred thirty millimeters in order to make it far more manageable to manufacture. The sole issue that I can identify is the corner on the one on the corner on the two. However, it is feasible to incorporate a radius into the equation. We can also select our fonts here once we have a 3D shape. I will simply see what this looks like in bold to see if we want to set bones or italic. It actually looks quite nice. The gap fruit should be sufficiently large to fit into the milling cutter, resulting in some being left as containers. I am somewhat fond of the appearance, and I intend to maintain the Ariel font. Therefore, we have concluded that matter. We have the option to select "okay" and examine the location. However, this does not necessarily indicate that it is permanently dead. I am currently examining this and can observe that the distance between the two and the edge is greater than the space between the one in the edge. I am suggesting that you perform a double-click on this feature. Furthermore, it allows us to revise it once more, allowing us to simply drag this point to reduce the distance, which appears to be approximately the same. Given that the origin point is located here, it is difficult to determine. However, the upper portions may be utilized, resulting in one two-and-a-half-square. I have approximately two squares, which is perhaps slightly more than I originally thought. Something similar, and then click on. All right. Now, that appears to be approximately accurate. It appears to be approximately equal. Subsequently, the 21 have been secured. We must now extrude Al once more in order to create chains with 3D objects. Consequently, we conclude our illustration by selecting here. We are going to rotate in order to gain a more comprehensive understanding of the situation. We can now select this feature and select "X street nice" before we can either bring it up or bring it down and remove it. Therefore, I will elevate us by 10 millimeters, which is equivalent to the depth of our side step. I am referring to the distance box, which is 10. Please press the return key. And now that we have a solid twenty-one on our key, it is flush with the top as we machine it. The only thing I desire to do is to inspect the corners, as we are unable to machine an ace with a round cutter. Regrettably, we have no other option, and any round cuts will interfere with his analysis. Therefore, this feature, this line, and all of the aforementioned lines require a boost, correct? Let us examine the matter. In order to accomplish this, we can invoke a sketch design shorthand, such as pressing the S key and typing "fillets," as we had previously intended to liberate the fillets. I will select this line. We can now extract that information and transform it into a fit of fury. Certainly, you have stated that now, which may appear somewhat improbable. Ensure that it appears satisfactory in the context of our broader objectives. Therefore, a five millimeter radius would necessitate the use of a five millimeter instrument to complete the task, or a lesser one. I will convert this to five millimeters and hopefully strike him with ten in this compartment. I would replicate this process for the two, as we now have a convenient radius that I can enter. I will make an effort to preserve it. It is five millimeters or larger, if feasible. So, in essence, press to bring up our show. It cannot be accommodated again. Click on "finished" and then click on the line in the upper right corner. I will now bring it out. It is a pleasant radius. We are now introducing this line here as well, as we have a pointed corner that is not easily machined. Therefore, it is feasible to exceed five millimeters this time. I propose that we increase the size to six millimeters. Therefore, we are discussing six and will revert to the same size fantasy. Another shortcut key, as I am capable of completing it. Choose "fit it" and then "select the vertices" to remove it. That would be five millimeters. Whoa, that is an excessive amount. Therefore, we are discussing a value of five to ten. Please observe the current state of the matter. Therefore, that is an additional machine that is required to complete laser 21. Therefore, there are already two machines. I am currently examining it. Our cutting machine is on the brink of entering the area and generating the letters. Consequently, the fundamental components of our key have been established. While we are at it, it may be beneficial for me to demonstrate engraving on the scene. I will see you out here. Royalties are a frequently employed feature. Therefore, we will compose a birthday greeting and a tribute to your long hair. Therefore, we once more press the s key to activate our design shortcuts and enter text. I will browse around to see if we can fit in "Happy Birthday" along here, and the first few letters will divide our working plain, which should be the face of the component we wish to engrave. Therefore, we will enter "Happy Birthday" into the text field. The text is a touch upside down, so we can rotate it by 180 degrees using this tool. We will also increase the size of the text. We have observed that the measurement is one hundred fifty millimeters. Therefore, I will adjust it to approximately forty millimeters and observe the outcome. Oh, it is nearly flawless. Observe that. In fact, I would argue that the present moment is ideal. That was an accurate prediction. Therefore, clearance is unnecessary at this time, as engraving is the case. Therefore, we will employ a sharp instrument, such as a sensor drill, if we do not have sand engraving pieces and the machine is capable of constructing these letters. Therefore, I will simply select "returns locked down in place." Ariel, the funds are retained by them; however, we have a variety of alternatives available to experiment with the funds in order to create additional scripts. Any phone that is installed on your device can be engraved, so we can leave it as Ariel for the time being and select on. OK, the thickness of these letters will not be as high as it is now unless we engrave them at a very deep level, as engraving tools typically operate at a 60-degree angle or, on occasion, a 90-degree angle. By placing a flat spot on the point of the tool, we can truncate the bottom, resulting in a wider lettering. However, the lettering will become increasingly wide as we go deeper. To enhance its visibility, we can also fill in the area with Indian ink or permanent marker. Therefore, these are a variety of suggestions and recommendations for engraving on items of this nature. Therefore, we are presuming that the lettering will not be this fabricated unless we delve approximately six millimeters.

At present, we have only 10 millimeters of material remaining to engrave grapefruit, and we will not be going as deep as that. I will simply conclude the illustration here and examine the final product after further consideration, as we will be lowering it. I will proceed to extrude once more, and we will select the lettering located here. Subsequently, we will have the option to extrude it either downward or upward. I mean, we could elevate it, but we would have to use a very fine cut to ensure that it fits between the letters. This will present us with complications when we attempt to perform two breaths. Therefore, if we were to subtract this by two to three millimeters, as the lettering is quite large, and observe the resulting appearance. Oh, I have wandered in the incorrect direction. That is not a concern; it does not imply that you should return to the previous state. Click on the lettering once more and this time, lower it. Therefore, we are considering a negative number. To make it negative three millimeters this time, we will subtract the free return. Once we have said yes, I would undoubtedly suggest a large truncated instrument. Therefore, we have a level area at the toe's end, or in a sense, Drew, as geos are typically around 60 degrees. Thus, these would be the ideal syllables for the slope of these letters. We should likely conduct a quick test run on a spare piece of material before beginning to engrave it to ensure that the lettering is satisfactory. This ensures that the actual components are not damaged during the engraving process. Additionally, it is important to maintain the sense of point of each letter. Fortunately, the word "birthday" does not contain any tiny, rounded segments in the midsection. Therefore, we should be in good health. We will now proceed to the subsequent lesson, in which we will examine the manufacturing of two components and the conversion of the program into G code that our sensing devices can interpret.

ROUGH MACHINING THE KEY

We have now completed the development of our model. We must commence the examination of two components at this time. It is a particularly advantageous moment to preserve it. In reality, we should have saved it multiple times prior to this moment. Therefore, to conserve. We have reached the icon located above the save button and have assigned it a version description. Sonja is a version 1. We have received the complete name and will retain it. In the event that we require any of our previous programs, we can select them by clicking on the later panel located above. All right. Therefore, we concealed the X in order to obtain a close-up. Perhaps we should examine an instrument and determine the method by which we will manufacture it. I would suggest that we should first machine out or this area, and then we can consider doing the engraving. Finally, the profile is cut to order a machine and begin machining the internal components at a rapid pace. This will ensure the highest level of rigidity during the machining process. Therefore, it is imperative that we consider the possibility of retaining our work in this location. Therefore, if we envision a sheet of wood with this in the center, we can erect clamps around the perimeter of the wood, assuming that the clamps are 15 millimeters in height. Therefore, in order to prevent any restraints from being struck while we are encircling the components, we will modify our rapid movements to exceed that limit. So, let's examine the beginning of the manufacturing process for these two components. We are hopeful that our radius is sufficient to accommodate our tools and that we can produce the part as we have designed it. To initiate the manufacturing process, we must exit our design area and proceed to the manufacturing area. To do so, we click on the box above and descend to the manufacturing area. This is the location of our encampment. Our computer-aided manufacturing process. It is the location where we operate our tool compressors. Consequently, the majority of this, if not the entirety, will be allocated to these machining processes. If we incorporate a profile with curves and ridges. We would then be considering a leisure day, but we will opt for 2D machining because the majority of the work you have done since you began your job is in fact 2D. However, prior to proceeding, it is imperative that we establish our billets in order to accomplish this.

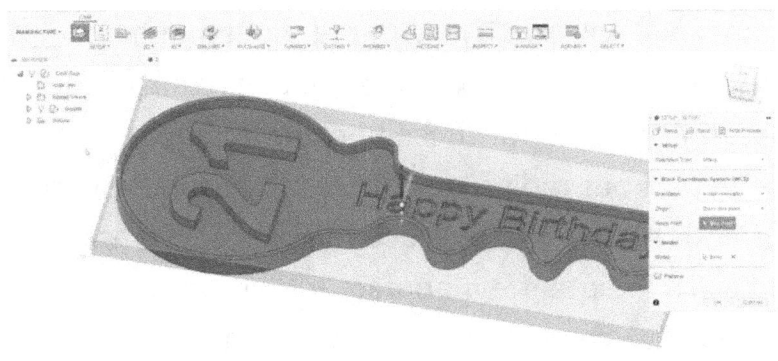

Clicking on "setup here" allows us to perform tasks such as adjusting the quantity of material on the machine table, which is classified as a 360-degree stock infusion. Stock is the term used to describe it. Initially, we will prioritize the million taps on our setup box operation. We can elect for milling on this one, as the alternatives are turning or cutting. If projectile discharges or laser plasma are employed, cutting is also an option. We may be examining a rotary router. Therefore, we will maintain it as a milling operation, as Sarah is essentially a standard. We do not need to stress about making too many changes to these crates; fusing 360 is quite effective out of the box. We can determine what we require and what is most beneficial for us. All right. Thus, we can determine the quantity of material that is excessive from our parts by navigating to the stock tab. Therefore, we would measure the material that we intend to place into the table and incorporate it into this process. Subsequently, we acquired stock options of varying sizes. I will now add a millimeter to the sides and the top. Therefore, we have an individual who is responsible for skimming off the top, and many individuals have nothing to add to the bottom. If we round up, there is nothing left. Yes. Therefore, this is the accepted practice. Once more, I will refrain from making any modifications to this. I proceed to post-processing. It is at this point that we provide it with a program that generates thousands of keys and is free of charge. We refer to it as the birthday key. Just to ensure that we are aware when we are reviewing our programs. Which program is which? There is no need to modify anything else in this section. Click on the noises. All right. and that is our responsibility. Establish a. We will now proceed to our initial 2D operation, which involves the removal of the material within our key. Therefore, our initial course of action is to implement 2D adaptive cleansing. This is an excellent method for roughing out material, as it adds a constant burden and is the most efficient method of removing material. Additionally, a 2D pocket option would be suitable, as it would allow us to route out our pockets and the images as we mouse over them. Provides a comprehensive explanation of the functions of each. Since I have not addressed any topics in this course, it is straightforward to determine the purpose of the operations and address them. Therefore, we can skim along the surface by employing a large cutter. I will proceed with the assumption that our material has already reached the final dimension. If required, we can always turn this program off at this point and before we execute it. Therefore, if we produce a few of these and certain components of our material are bulging or not flat, we can quickly add a face-facing operation and then execute the second program to create the key to the console. This will likely be used for the exterior profile. We can perform this operation at various levels to avoid cutting the entire piece at once, and we can also add finishing cuts and finishing passes. Consequently, I prefer to employ adaptive cleansing for internal projects whenever feasible. It is the most effective method of removing material. Additionally, when conducting this activity in an industrial setting, it is imperative to prioritize velocity. We are interested in expediting the release of these products, with a particular emphasis on the hobbyist market. We are merely interested in producing the most desirable components. However, in the industrial sector, it is imperative that our programs operate at the highest possible efficiency in order to optimize the revenue generated by our machines during the machining process. Therefore, I will opt for 2D adaptive cleansing. The initial step is to select. Therefore, it is evident that these radiuses are five millimeters in size. Therefore, it is a negligible quantity of material to eliminate a five-millimeter cutter. Therefore, I am considering the possibility of performing a preliminary cut with a larger blade and subsequently utilizing a 5 millimeter blade. Therefore, we should implement this approach and observe its effectiveness. We can always reverse it at a later time and complete the task with a 5 millimeter blade if additional guidance is required. I will likely choose a size of 12 millimeters or 10 millimeters, or something similar. Therefore, we were a form of tool if we ascended to the summit. I am going to select a flat milling implement and proceed quickly. All right. Now, we must assign it a dimension. To do so, click on the "dimensions" tab and select the "diameter" option, as this is the time of interest. We have the ability to modify any other information in this section to correspond with the instruments in our toolbox. However, we will proceed with a total of half an inch. I will be a twelve-point seven millimeter instrument, as twelve-point seven is equal to one-half of twenty-five point four. High-speed operation continues. That is acceptable, provided that you possess carbide or ceramic. This primarily influences the recommended rates and inputs that the machine will provide, as well as the number of fleets.

For the time being, we will presume that all of this is accurate. However, you can configure this to use the tools that are available to you, and Fusion will recommend the optimal velocities and feeds for the total. We have now selected the desired diameter. By clicking on "all," we will be presented with a comprehensive inventory of all the tools available in Fusion to create our part. You are now able to create your own tool library using the tools that are currently available to you. Therefore, Fusion is aware of the resources at your disposal, and you can immediately select from the list. Therefore, this is a project that should be developed gradually. Therefore, inform the software of the precise contents of your personal inventory. Therefore, we have a suitable option for bronze aluminum and only select aluminum. The cutting angles are strikingly similar. Therefore, aluminum is suitable for our requirements. We do not have any wood in particular, but we do have plastic, which is also very similar. We opt for aluminum, which has a flat bottom and a total thickness of twelve point seven millimeters, or approximately half an inch. I will simply click "okay" to select this tool, and it will display a visually appealing 3D image of the tool to allow us to see how it will function. I will leave all other details as they are. Fusion is adept at determining the appropriate rates and inputs for us, as space and phase are vast subjects. Therefore, I prefer to allow Fusion to make the decision. If the cutting process feels excessively rapid, we can either reduce the speed on the machine or return to Fusion to modify our velocities and flows. So if we select this outside profile in the bottom corner, Fusion will recognize the bottom top half that we wish to select. This is in contrast to the numbers that are coming down from the stock consoles, which are related to removing stock. However, we are not working with a stock hex; we are focusing solely on the models, so we can skip that step. Therefore, we will not be employing rest machining at this time. We will utilize this when we arrive with a lesser instrument. This function exclusively eliminates content that has not been previously eliminated. Therefore, if we introduce a 5 millimeter radius to the radius in this area, it will only machine these races and not the entire surface. Therefore, we will employ the rest machine for the subsequent operation, but not for this Raptor run. We do not need to concern about the spigot rising up in the middle, as it would require machining around it. Currently, the direction in which R2 is coming in is the orientation. Our instrument will descend along the z-axis, rather than from the side. We would modify it accordingly if it were approaching from the side. However, this pertains to multi-axis machines, including those with five or seven axes. Therefore, the geometry tab is transitioning to the heights tab. We are now going to presume that we are using 15 millimeter high clamps to secure the outside in an outward direction. Additionally, we will require an increase in the upper clearance, which is currently 15 millimeters. Therefore, this is the method by which we modify this. We rotate and maneuver to enhance our visibility, and we are currently examining the clearance height. Therefore, we can simply elevate this and increase its thickness by millimeters. Our clients have provided us with a 50 millimeter clearance above the material's retractable limit. I am going to make this 20 millimeters in order to ensure that we do not encounter any restraints. Therefore, we can simply discuss 20 millimeters here and 10 millimeters there, and that will be the addition. It is now acceptable to return. It is currently resolving the two components, as it is evident that the corners are not quite captured. However, this is acceptable, as we can use a different tool to eliminate them at a later time. The sparring effect of these tools is evident. This is the method by which the material is being removed. This is the reason we employ an adaptive 2D work environment that is integrated with these tools. It operates under constant pressure, consistently removing the appropriate quantity of material, and remains under the same burden. It is remarkable that no errors have been detected, as this is the most efficient method of machining. Therefore, we are confident that this manufacturing procedure will be successful. Occasionally, we may encounter an exclamation mark, which we select, and it reveals a collection of dwellings. What we have done incorrectly, in order to return and rectify the situation. So, if this tool is configured in this manner, we can now run the simulation by selecting "Simulation" and then "Play." We can observe the tool removing the material, and each time we rotate the part, it pauses, allowing us to observe that it is performing this in a single cut, scaling down and going round. We could now perform this and two additional cuts, and then provide a report in a moment. Additionally, we could arrange an additional cut to demonstrate how it would function. Therefore, we can incorporate additional machining operations into this after our instrument has passed. Nothing is permanently fixed. We can always revisit and make adjustments to the work we have already completed. If it takes a while or if we wish to expedite the process, we can move a slider up. We can then observe the machine and determine that there will be no collisions and that everything is in the desired state. The yellow lines in this case indicate a swift movement.

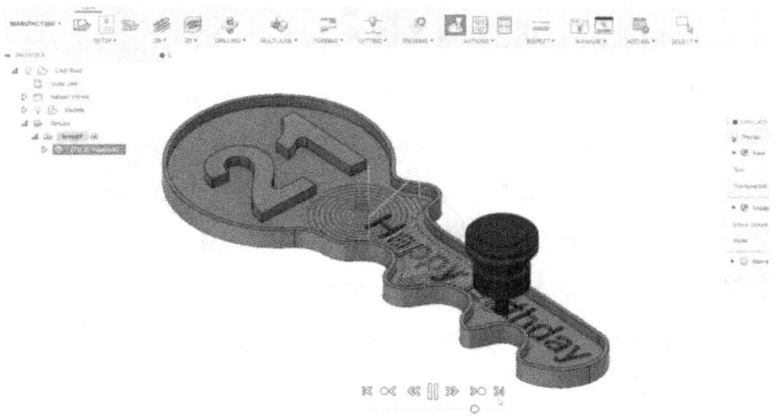

This is the point at which it transitions from the task. We have just adjusted these to 50 millimeters above. It is 50 millimeters above the plane, as you can observe. All right. Now, if we were to do this in two cars instead of one, we would open our menu here. We can double-click on the first item to modify it, and we can double-click on the second item to return to our box, allowing us to make edits. Once we have completed the height adjustments, we will hit the return key, which will initiate the assimilation mode and the coding of a program. We neglected to configure the passes. Consequently, we can proceed to the subsequent menu and select this option, which will allow us to determine the number of passes we wish to execute. We can now conclude with the fifth topic. Please halt. Clarence, this is intended for the final edit. That is acceptable. I will refrain from addressing that matter. We are seeking to complete the task and make the final cut with multiple fatalities when we arrive to handle the lesser agencies with the smaller toll. This is because we wish to take more than one cut. Therefore, we intend to execute two or three cuts in order to achieve the utmost roughing step over a set of 10 millimeters. Now, if you recall that the distance between the upper surface of this object and the bottom surface is 10 millimeters. This is the reason it is completing the task in a single run. Therefore, if we adjust that to five millimeters, we will observe that it completes the task in two cycles. The direction of our cut ahead can be altered to climb Milin or conventional. I will leave it to the climate in McNair, as the machine is rigid enough for this process. If you are using a machine that is not rigid, you may switch to conventional. The only way to determine if the machine is giving a bad charted finish or is not as you desire is to make a test. It's probably due to the climb Millin where it's pushing back on the bearings on the lead screws some expensive machines may have reoccurring lead screws which have ball bearings around the internal fret so there is no backlash but a lot of machines won't have this feature so very high end feature so bear that in mind we might might need to use conventional melon if your machine doesn't have a reoccurring both leads screw Okay so change in what we did there I am now going to run another simulation to see if this does this and two cuts so we can click on Okay I'm going to let the tool pass build you see here the percentage going up this is showing you how it's calculated needs to pass and how long it's gonna take to work out how to remove all this material and when we're done we can click on simulate and play see how this works I'm going to speed this up again if I just pause that for a minute we can see these blue lines actually off the surface of the material it's hard to see maybe we can zoom in Yeah you can see it's actually taken a five millimeter depth cut there and it's still five millimeters below it for our second pass so that works. So let's do this in two cuts. Go ahead and play again, and observe the machine work its magic. Once it's finished, the first cut poses again. The reason I say this is that it's now going into the bottom cut is that it's creating more blue lines underneath this one. You can almost see the cost of breaking free here. Therefore, this is our second set of edits. Therefore, we are executing this in two edits, as per our specifications. Therefore, we will reduce the speed and engage in playback to observe its functionality. Manually dividing that program into two components would require an immense amount of effort. This is where Fusion 360 truly shines, as it provides us with these adaptive tools. It is possible that it will generate a significant number of lines of code to accomplish this. However, there is no need for concern, as modern computers require RAM to function effectively. I recall the first time I began operating a machine, which was a long time ago. We were required to exercise extreme caution regarding the duration of our programs, as the devices had a fixed amount of RAM of less than 1 megabyte. Typically, a few critical bytes are required; however, RAM is no longer a concern in the context of contemporary computers and machinery. Therefore, the scale of our programs is no longer an issue. Additionally, this evening is enjoyable. Why was g 0 1 Shorten's 2 g 1? This is to generate the quantities of bytes that were utilized, or to reduce the amount of bytes that were used in our programs in order to keep them concise. We are currently experiencing it at an accelerated pace. They are returning to complete the task, and it will perform a final pass at a slower pace to ensure that all details are smoothed out. This process involves taking free cuts. The previous iteration was simply returning to complete various features that were not completed during the initial run. Therefore, I am exceedingly satisfied with the process. I believe that will be an excellent decision. Provide us with an exceptional conclusion. It will remove the material with great efficiency, and I will save the program there before we begin working on any additional tool parts in the event that any errors are made. We are aware that this is acceptable at this time. Someone is required to rescue. Give it a different version and Bangor quiz 1 v2 and select "okay."

FINISH MACHINING THE KEY

Servers are currently in the process of roughing it. We made two distinct cutbacks to reduce our food debt by 10 millimeters and additionally by half an inch. We will now examine the second pass, which is our finish and pass. This pass will also be used to conclude these competitions with a reduced toe. I will be employing a radius of five millimeters. I will employ a four millimeter tool to manufacture these races, rather than utilizing the tool's diameter to carve them. This will result in a superior surface polish and high accuracy. Consequently, the process of generating a radius is not the same as the process of creating a radius with a cutter. The cutter's diameter is measured and the cutter is pushed into the material, causing a radius to form. To remove material, we use a geo two or geo free command to move the cutter around the radius and remove it, rather than pressing it into the rodents. So we return to our 2D operation, and the timing of using a 4 millimeter is once again cleared. Some individuals select the tone, alter the diameter of our skulls to 4 millimeters, and click "okay."

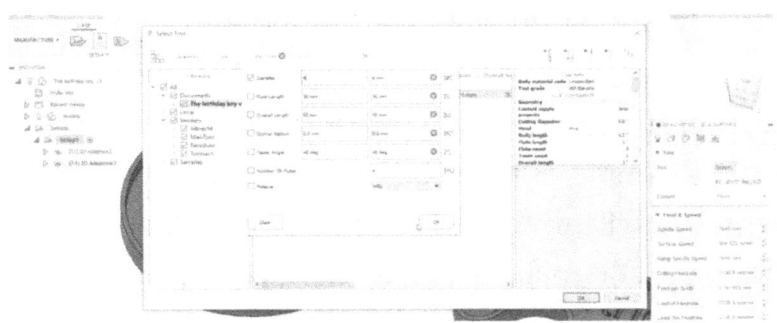

Currently, when we select the previous menu, we are presented with a variety of options. We are no longer employing the instrument, as we do not wish to drill. We desire a flat end, not a globe in the center. Melissa, this is a male flat end with a four millimeter diameter. It is not flawless. Slept that and an Okay now, during which we employed a four-millimeter instrument. It is feasible to envision the duration of time required to eliminate all of this material; however, it accounts for only two. However, we can simply eliminate the features by utilizing it in the manner in which we are currently doing so. That is logical; however, it did not alleviate the situation. Therefore, we will proceed with the same selection of features as we did previously. We are interested in the bottom area surrounding our numbers and will be examining the rest machine at this time. The remainder of the machining process involves the removal of all components. The other two did not, and only those components were affected. Therefore, we should also inform them of their existence -- a sample of MMS before we proceed to our aspirations. We will now need to remove our clearance height hair up to 40 millimeters, as we did previously, to ensure that we have the clearance and our stock diameter hair stock offset. We will also advance that. Consequently, you select that. I received 20 MMS and recommended the highest clearance because it contributes to the allure height. That is acceptable, provided that the shackles are absent. I am not overly concerned with the addition of this feature, as long as we have a Z-axis movement of 50 millimeters, which is just under 2 inches, on our machine. That is all that matters to me. We should be able to proceed without issue; however, we will need to ensure that the machine does not halt while it is moving the cutback, as we are requesting that it move a greater amount than the machine's capacity. Therefore, it is imperative that we maintain a state of equilibrium when establishing this. We are not leaving any stock this time, so to remove stock to leave now means we are performing a finishing pass. Everything in the house appears to be in order, so I will click "okay." We will observe whether this calculation is cutting puffs and whether it will be effective. She has calculated our and past now, and it will determine where the two can enter to remove material that is larger than the current size. However, it appears that the issue area is only the letter and number two, which we anticipated. We still have 20% of the material to remove. And yes, that is acceptable; Celeste will always go in and eliminate a small amount of material that was left on and around the two hairs.

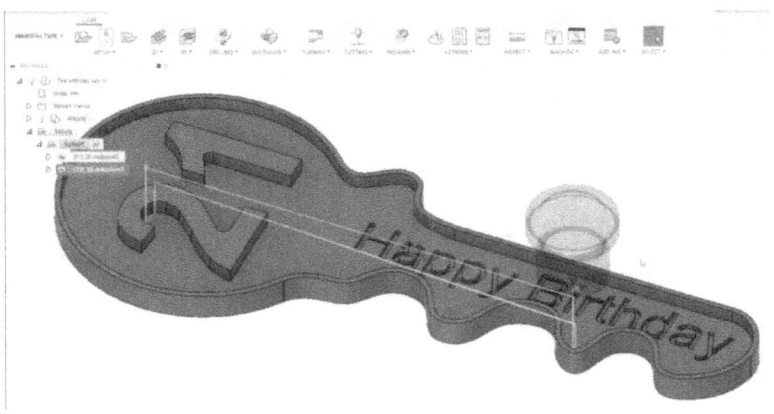

We have now removed a total of 4 millimeters of material that the larger instrument was unable to remove. Now, if you recall, we left the concluding pass on the large crevice. I will now return to complete our finish, in part because we left a half-millimeter stock material on all sides to accommodate a finishing pass. Therefore, the following days, you will be required to apply our finishing cuts to the primary internal components. This will require a return to 2D and 2D adaptive clearance. Upon returning to our half-inch purchase, I recall that twelve point seven is represented by a click and a K. Subsequently, we possess it, as well as this document. This is the same tool that we previously employed, and it has been added to our document to include the birthday key. So, by simply double-clicking on that, we have selected our half-page to perform our finishing pass. They may wish to use a different tone, such as a different number of TIF finishes, but we will be using the same tool that we used to rack the materials, which is quite delicate. So, it is not as though we will be able to effortlessly reconcile the two. There is no deviation from the norm. We must first identify the features that we wish to complete the machining process for, and then proceed to the heights. This would necessitate another modification. Therefore, our clearance site has the capacity to accumulate a minimum of 50 millimeters and counterbalance it by 20 millimeters this location. In the same manner as we did previously, we will now advance to permits. We are currently unable to leave any stock, so we will be turning it off. Therefore, this is prior to our departure from a half-millimeter margin on all sides. Turning this off will allow us to machine it to the appropriate dimensions. This is an option that must be completed and approved. If you are achieving a satisfactory surface finish while removing the material with the roughing tool, it is beneficial to perform a finishing pass for practice and improvement. The call is seamless and pleasant due to the presence of this moving operation. Therefore, we do not observe any angular openings with a high turnout on it. We can establish a move intolerance of naught, which is an exceedingly small value, rather than one millimeter. That is the temperature for the thickness of a human hair or something. Therefore, we can modify the statement that states there is no point five. I will simply select "okay" and observe the current situation. The instrument is currently in the process of constructing an impasse, and it will be removed from the surface by half a millimeter. We reduced the size of the tool when we entered with our compact four-millimeter tool. We did not actually leave any material on the premises. So, as long as we have configured our tools appropriately, this will seamlessly integrate with that tool. Flip to assimilate and select "play." I am aware that it has completed and exited the bottom, and all edges, faces, and components have arisen from the ground. Therefore, we have completed the pass. Now, we must proceed to the engraving. All right. Now, in order to proceed to engraving, we must navigate to a 2D operation. Engrave is available in this section, and we can select it to investigate our engraving procedure. It is providing us with an additional operation at the bottom of the list, allowing us to review the operations and make any necessary adjustments. It is always possible to return and make changes. This is the reason why Fusion 360 is such an exceptional piece of software to operate. All right. We will now examine the engraving of an instrument. We have a variety of options. If you possess an engraving tool, it is exceptional; however, a sense drill is also an effective engraving tool, as the angles are precisely suited to our requirements. Therefore, we should determine whether we have any engraving tools in stock or if we can utilize a sensor drill to disable the diameter hair. It is possible that a sensor drill would provide us with straight sides for our letters, while a spotter would provide us with sloped sides. The angle would be contingent upon the drill, but I am aware that we prefer the engraving of a sensor drill. Therefore, we can simply click "clear" and "okay" to navigate to the tool type menu.

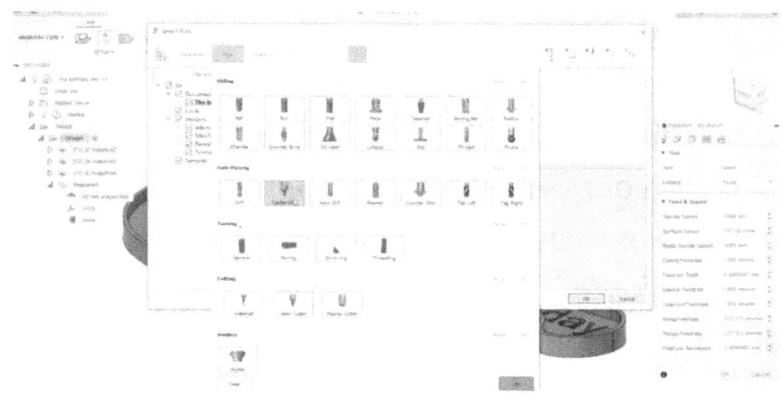

Therefore, we will execute the flex and continue to click "okay," as well as potentially selecting "number two" or "number three," as drill would be an ideal solution for this. Therefore, there is no freebie, as Joe has a quarter-inch diameter on the faucet. Therefore, we will select this item and click on Ike. We must exercise caution in this situation, as the margins of our work holder will collide with the size of our material due to the fact that Drew is not drawn out sufficiently. Okay, that is acceptable. We have the ability to modify that within our arsenal. Therefore, if we initiate the call and prolong the sense drill, and if we do not have sufficient clearance, we can always employ the spot drill. That is significantly lengthier. Additionally, it is feasible to lengthen its length. Therefore, we should determine whether it is possible to modify this to indicate that we have our sense drill on hand and that we can proceed to construct our own instrument. Therefore, that is what we will do. That serves as an effective illustration of the process by which we can establish our own sensor drill. Subsequently, we acquired a cordless sensor drill at no cost. This is a sensible choice for this long series, as we will require a sensor drill to be present on site for an extended period. We will reduce its length and provide additional information to determine the body length or shaft diameter. Anything that we desire may be altered. Therefore, it is the body that is inert. I am intriguing. I am interested in deepening my understanding of the matter. The current time is 6:35 p.m., or a quarter of an inch. The initial step is to modify the type. So we could modify our type to a sensor drill. Here, we would change the tip diameter to six point forty-five millimeters, which is the standard for number three sensor true. The upper body length of the hair is 50 millimeters, but we only require 10 millimeters plus clearance. Therefore, we could make the number of flutes on these two items 20 millimeters. Therefore, we can reduce that to two, and everything else appears to be in order. So, it is evident that there is an issue with our body, as evidenced by the remaining hairs. The body left must be larger than the shoulder length, as indicated by the red writing that appears when we hover over it. Therefore, that is not an issue; we are aware that our shoulder is obese millimeters. So we can adjust this to

40 millimeters and press a 10. This is then viewed in the "number of flutes" section, which has been reset for some unknown reason. We can now set it back to two and select "Okay," which will return the screen to a blue color. So that is how we configured our own within Fusion 360. We can now press "okay" and select the two. This provides us with a greater sense of clarity from our tools, which inform us that we can enter the area directly. This was engraved without incident. It will not collide with any of our phalanges. All right. Therefore, we must now select our geometry after selecting our two options. Subsequently, select the dramaturgy tab. The objective is to focus in by selecting the lower portion of the letters, rather than the upper portion. Select all of the geometry and the lower right of the letters. Therefore, the machine is aware of the extent to which we must drill in order to remove these. Currently, we are interested in conducting a small amount of experimentation to ensure that we achieve a satisfactory match. Consequently, it is advisable to send this portion of the test plate through the engraving process to ensure that it is produced in the desired manner before we cease cutting on the main part. It is primarily due to the tool's angles, which we have chosen and the manner in which it will function with this. We also need to select the inner geometry of these, such as the portions of the O's, the A's, and the B, as well as the final an over here, as in Excel and PS. Greetings. Okay.

We will now simply click on our heights to ensure the safety of our retractable roof. For some reason, any grass that we retract is acceptable, as we must ensure that it is above 50 millimeters. For example, I can raise this retractable roof to that height, and the other retractable roofs will be 10 to 20 millimeters higher once we lower the clearance one. Therefore, as long as the clamps are concerned and our clients are 15 millimeters away, a machine can accommodate this type of travel. Subsequent passages will be made with an acute angle corner. That is essentially equivalent to our toe. This is the appearance that our letters will have on the sides, as they may be inclined at an angle similar to that of a toe. So once that is completed, we can simply move on. All right. We are experiencing an error silence. It is necessary to modify our instrument, and I anticipate that this may occur. Therefore, we can return to the area and reconstruct the opening. Therefore, I will employ a standard spot drill for this task, rather than a sensor drill, which is acceptable. For this reason, we must proceed to the type and spot drill section and determine the inventory available. Okay, a spot drill with two flutes, which is equivalent to six point forty-five millimeters. That will suffice, and we should have sufficient clearance to overlook the component's dimensions. We should inspect the area to ensure that there is sufficient clearance. The measurement exceeds 10 millimeters, which indicates that there is pressure. Then, determine whether this is feasible. All right. Our aspirations for swift clearance are significantly greater than those of the clamps. Therefore, that is acceptable. We should conduct a simulation to determine the functionality of this. We simply distribute the cups evenly. If you determine that you may require multiple incisions, we can incorporate them into the toolbox. I believe that one cup will suffice for this task after we have chosen our simulation. Therefore, that is the course of action we will take. There are no errors. Everything functioned as intended.

PROFILE ROUTING AND POST PROCESSING

We are currently in the final stages of manufacturing her case. We will only be conducting the external profile, so we will be using the festival for that. This time, we will be examining the [REMOVED] console for two days. Our objective is to complete the exterior profile in a few runs, allowing us to select the console. So, we will select the half-inch and mil or twelve-point-seven-mm end mill that was used to complete the internal part of this product. This will allow us to double-click on the tool to select it and reopen the 2D control box. So, as with the previous geometry, we will need to select the bottom edge of the outside profile to select our geometry. We do not want to leave any material on. Therefore, that is acceptable; however, we would prefer to maintain the faucets. It is entirely a profession. During the machining process, I disappear, and we can subsequently eliminate these types by using a file and sandpaper to clear up the areas where the taps are located. This will prevent us from selecting taps. Now, it is unnecessary for them to be in close proximity. We do not desire an excessive number of touches. Alternatively, it would require a significant amount of effort to eliminate them later. We can maintain optimism if it is unfettered, Pt. Five and one.

That is exceedingly precise. However, the free millimeter border is acceptable. Once the part is machined, we should be able to remove it by hand. However, if he locates the parts, it will be difficult to remove them the second time. When the machining is completed for the upper position, we can minimize this feature to make it significantly simpler to detach laptops by hand. Based on the distance, it is 100 millimeters. His ohms just stated that federal power. I am going to make this one hundred and forty millimeters. We do not require a significant amount of it to be present in this location, as it will only be removed from our plywood panels or whatever we are cutting during the final cut. Now that our faucets have been completed, we can proceed to our hypes. As before, we must increase our clearance aspirations by a bar of the clamps, which is 15 millimeters. Additionally, we must adjust our stock to 20 millimeters. Therefore, we have an abundance of clearance; however, these must be precise, as I have previously stated. We are only concerned with the absence of our restraints. All right. I will proceed with this in a few passages, as we are removing 20 millimeters or 15 millimeters of material in two passes.

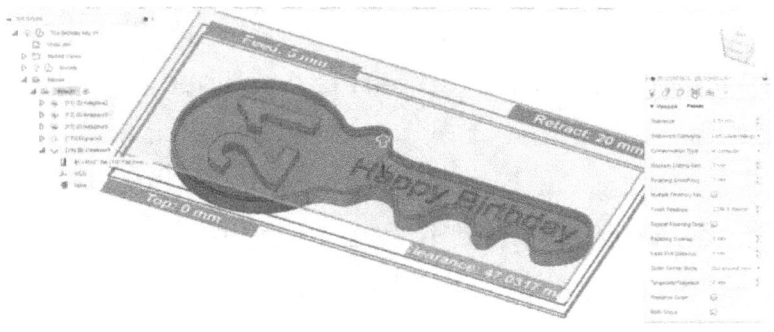

Therefore, we will execute this in a series of runs, utilizing multiple decks that are capable of roughing out individuals without the necessity for a rough cut. I will conclude the final demise, and we can also choose to complete at the final depth half. This implies that we will add a small amount of material to the sides of the final tool path in order to improve its appearance. We do not desire to leave any stocks as a concluding cut, but we do need to address our step-downs. The millimeters were a touch tiny; perhaps 5 millimeters would be a slightly better option. We will be making five millimeter deep incisions with a half inch to our utmost roughing step. The next step is one millimeter, and I will repeat this process until I reach five millimeters. I wish to make multiple finishing passes. Ascend. Therefore, if your machine is capable of accommodating it and all other aspects of the situation are satisfactory, you may modify this to conventional military if necessary. I will simply select "okay" and observe the appearance of this. Now, we have the option to return and modify the process if we are dissatisfied with the results. As evidenced by the five cuts and five passes, which are equivalent to five millimeter-sized portion cuts, the final cuts will circle and remove a small amount of the sides to achieve a finished look. Therefore, that appears to be satisfactory. We have no errors in our calculations, so we can simply run our simulation to observe the appearance of the object and ensure that there are no crashes or conflicts with other objects. We can rotate the object as we are doing it at the end, where it pops out and returns to a different definition. The final pass will be performed as it descends, and we will be cutting the tops out with another photo pass. The tabs will be slightly materialized during this pass, so they will not fly out during the cutting process. We have completed our portion of the task. We can now upload this into our machine after we have processed it through the postal processor and generated the program Phyllis. This machine is capable of reading it, and we can cut it on our Walter. Before we conclude this lesson, let us visit the postal processor to learn how to convert this into a program. Therefore, I can select the post processor from this list. We have already provided our program number and comments that are unnecessary; therefore, the only task remaining is to determine which machine we have brought into town.

For example, if we open this and select our machine from the drop-down menu, it will convert a program into G codes that our machine can read. This is because G codes are the same, but each machine has its own unique preferences. This ensures that it is fully compatible with your machine, as some machines utilize distinct M codes. Therefore, we should examine the devices that we possess. We have a since he composed the components using a Mac that is gratis. Therefore, in the event that we possess Haydn Hame, which is a distinct language from G code, that could be a viable alternative for our machine. She typically conceals him behind large industrial milling equipment. It is strikingly similar to g code. It is devoid of G codes. It employs linear movements rather than g codes; however, this instruction will be saved for a subsequent day. That is an entirely distinct topic. Therefore, we acquire rolling machinery and a retail box. I am going to use a Mac that is free of charge, as it appears to be the standard for home rule to use someone's home. We can simply click on "post" and the program will be generated, which we will save on my desktop. Currently, brackets are displayed in brackets that are included with Fusion 360. Front-end developers, including web developers, frequently employ this code editor. However, it is an excellent method of displaying code. Currently, we can locate this item. The program is lengthy, with numerous operations; however, it provides all the necessary information for our machine. This format is consistent with the G code teachings found elsewhere in this course. Therefore, you can acquire the ability to comprehend this information from the other classes to ensure that you can create and modify these files as you see fit. Currently, it is possible to observe that there are five and a half thousand lines of code by scrolling to the bottom of the page. Therefore, there is a substantial amount of code that is prepared to be loaded into our machine. I believe that the final step is to proceed to M9 to deactivate the refrigerant, and an emphatic message on the machine indicates that the program has concluded. Therefore, I am content with the outcome. Consequently, we can close the brackets and save the file in the finalized state (code). Now, all of our files have been saved. Press the return key. Therefore, if we wish to revisit and modify him at a later time, we can do so by filling in the form. This is how we complete the module with multiple tools, multiple features, and multiple methods of routing, including internal profiling, external profiling, and engraving. This concludes the extensive lecture on the use of Fusion 360 to create a comprehensive part.

INTRODUCTION

C c Turing machine. What is the definition of panoramic landscape C? It is a computer that operates on numerical values. CNC controllers direct the precise movement of the tool along multiple axes in the process of numerical control, which is the process by which a computer controls the operation of a machine tool. Perform tool adjustments and numerous pull-off features, such as coolant control, and regulate the cutting field, spindle speed, and rivulet. The utilization of C and C machines is on the rise.

> **Computer Numerical Control**
>
> Computer numerical control (CNC) is the process by which a computer controls the operation of a machine tool. CNC controllers direct the precise movement of a tool along several axes, regulate spindle speeds and cutting feeds, and perform tool changes and various on-off features such as coolant control.
>
> The increasing use of CNC machines is a continuing trend in manufacturing, due to the growth in demand of production output and quality improvement. Because of better engineering designs, advanced tooling, simplified setups, and larger tooling capacities, CNC machines have become faster, more reliable, and more efficient. The functions CNC machines perform make it easier and more profitable for companies to compete in the marketplace.

Manufacturing trends. The n C machines have become more reliable, efficient, and quicker as a result of the increase in production demand, as well as the quality enhancement that has occurred as a result of improved engineering designs, certain advanced tooling, simplified setup, and larger tool capacities. Subsequently, the functions of C and C devices facilitate and profitably enhance competition among businesses in the marketplace. Competition. Market.

ADVANTAGE

What are the advantages that we have observed? As he provides numerous advantages to the manufacturing industry, one of these advantages is the enhancement of safety. gorges or places in place to safeguard the so brittle, the operator is not as closely associated with the machining process. The machine controller is typically situated in a secure location to safeguard the operator from the cutting instrument and the machine's forward movement. A board program that is specifically designed to execute numerous specialized operations provides a greater degree of flexibility. There is an extensive array of machining operations that can be characterized as blue. Increased precision eliminates the need for an operator to operate the machine. As a result of movement, it is now possible to maintain closer tolerances for reduced board and achieve greater accuracy. Engine. On the path to reducing the preparation time. Accelerating production and employing pre-purchased software. Eliminates the necessity of maintaining a substantial inventory, as both can be manufactured in a shorter period of time. Reduced storage program deletion and a reduced number of fixtures enable tasks to be completed more rapidly. Reduce the cost of cutting tools, particularly those that are near specific tools. The machine can confirm the tool path to the tool at a lower cost, so it is no longer necessary to use the tool.

> **Advantages of CNC**
>
> CNC offers many benefits to manufacturing. These benefits include:
>
> - **Improved safety.** The operator is not as closely associated with the machining process. Guards are put in place to protect the operator. The machine controller is usually located in a safe position, protecting the operator from moving machine parts and the cutting tool.
> - **Greater flexibility.** With part programs designed to perform many specialized operations, a wide variety of machining operations can be employed.
> - **Greater accuracy.** Removal of the operator from controlling the machine movement has led to greater part accuracy and the ability to hold closer tolerances.
> - **Reduced parts inventories.** Reductions in setup time, increased speed of production, and the use of stored part programs eliminate the need to carry a large inventory, because parts can be produced in a shorter amount of time.
> - **Reduced lead time.** Stored programs and fewer fixtures allow jobs to be more quickly performed.
> - **Lower cutting-tool costs.** Specially ground form tools no longer must be used because the machine can generate a tool path to conform to that tool's shape.

Fixture that is situated lower. Coarse tasks that previously necessitated multiple fixtures. To generate the import number. In many cases, the number of fixtures is reduced, and only one is required to increase the life of the cutting tool. This can be achieved by increasing the speed and feed, or by simplifying the machining process to facilitate quicker production. There is a reduction. Operator-machine interaction. Decreased storage capacity. A reduction in the number of features and the number of boards in storage. Expands the amount of space that is available for alternative purposes. The machine's capacity to rotate in accordance with computer-generated instructions is intricate. In this application, the generated toolpath simplifies the reduction of complexity, which is facilitated by the increased accuracy and consistent speed. A reduced number of devices have the capacity to observe and continue to execute a diverse array of operations. Eliminate the necessity for numerous conventional machines. Decrease the duration of machine machining. It is compatible with and optimized by the high robot positioning. Handling gritty grid necessitates clipping inputs along the list. Significantly decreases the duration of the machining process. Scrap minimization. Reducing human errors is the primary objective of all machining accuracy. Your arrangement is a concern regarding travel limits and results in a reduction in refuse. Additionally, the allocation of systems facilitates the reduction of the number of configurations. The precision of production scheduling is enhanced by the use of controlled dimensioning, time, and manufacturing words. Reduced expenses. An aggregate cost savings is the result of all of the enumerated advantages.

TOOLS

C c rotating sintered tools. I was unable to locate the instrument that would allow me to select from a variety of options. During the process of learning your programs. It is important to note that the centerin tool is listed with only one number and is obtaining the actual measurement. It is certain that there will be additional locations that are not suitable for drilling instruments. Utilizing numbers as an instrument. A committee is a committee in accordance with this. Right here. The following are the instruments. Turning instruments that are designed for cutting.

1. Turning

Turning is the most common lathe machining operation. During the turning process, a cutting tool removes material from the outer diameter of a rotating workpiece. The main objective of turning is to reduce the workpiece diameter to the desired dimension. There are two types of turning operations, rough and finish.

Rough turning operation aims to machine a piece to within a predefined thickness, by removing the maximum amount of material in the shortest possible time, disregarding the accuracy and surface finish. Finish turning produces a smooth surface finish and the workpiece with final accurate dimensions.

Different sections of the turned parts may have different outer dimensions. The transition between the surfaces with two different diameters can have several topological features, namely step, taper, chamfer, and contour. To produce these features, multiple passes at a small radial depth of cut may be necessary.

Production of skins. With regard to the sight. Utilizing steel. Tool numbers for uneven shaping. Instrument. Three professional relocation tools. Completion. Instrument for turning. Complete the recording of the utility. Tool for transportation. Editing software that is conventional. Seventh. Threading. Twisted. Incorporated. Instrument for drilling. I will rotate it. Groove implement. Remove a portion. Finished. Tool for threading. Reamer. A device for applying tape.

TURNING

Next, we will discuss operations. Rotating. The least prevalent machining operation during the turning procedure is turning. Material is removed from the outer perimeter of rotating workings using a cutting implement. The primary goal of turning is to reduce the old wisdom to the desired dimension. The off-turning operation is designed to machine observe EVs within a predefined thickness by removing the maximum quantity of material in the minimum possible time. There are two categories of turning operations: cut off and finish. This documentation of the accuracy and surface finish finish read finish turning reduces a uniform surface finish. Our exam concludes with precise and precise dimensions, which are distinct from one another. The outer surfaces of certain sections of the rotating boards may differ. Dimensions of the exterior. The transition between the surface with two distinct diameters can exhibit a variety of topological features, including step tables, diversion for and count to go, and control to produce. Multiple compartments are included in this feature.

---

1. Turning

Turning is the most common lathe machining operation. During the turning process, a cutting tool removes material from the outer diameter of a rotating workpiece. The main objective of turning is to reduce the workpiece diameter to the desired dimension. There are two types of turning operations, rough and finish.

Rough turning operation aims to machine a piece to within a predefined thickness, by removing the maximum amount of material in the shortest possible time, disregarding the accuracy and surface finish. Finish turning produces a smooth surface finish and the workpiece with final accurate dimensions.

Different sections of the turned parts may have different outer dimensions. The transition between the surfaces with two different diameters can have several topological features, namely step, taper, chamfer, and contour. To produce these features, multiple passes at a small radial depth of cut may be necessary.

It may be necessary to increase the radial depth. And here is our exam, which pertains to the original diameter. The diameter was also reduced. The direction of the feed and the turning implement. Turn that is steep to steep. Generate two surfaces using the value of n. The diameters of the orbits vary. The final feature is the use of symbols to represent each phase. Tables. Rotating. The angle motion between the workpiece and the cutting tool reduces the uranium transition between the two surfaces with different diameters when the tables are turned. Displayed for rotation. Similar to the abrupt curve. Chamfer turning is a process that involves the creation of an angle transition between two surfaces, resulting in a solid square edge. The extent of the turn is determined by the turning down. Contour turning and contour turning. Operation. The instrument for carving. Axial motions. Both possess a highly defined geometry. In order to achieve the intended contours in the workpiece, it may be necessary to rebound the contouring tool multiple times. Nevertheless, the same contours can be generated using tools. The shape is that of a solitary vehicle.

FACEING AND GROOVING

Fronting. Machining the duration of the work is slightly longer than the final volume during the meeting. The operation of machining the end of the workpiece that is perpendicular to the rotor is known as "should be facing." To achieve the desirable part line and a uniform face surface, the rotating axis of the tool should be moved along the radius of the workpiece during the facing process. By removing a narrow layer of material. The cutting implement for the workpiece is located here. And the direction. Angelo. And now, let us discuss the art of swaying. Grooving is a turning operation that generates a narrow cut groove in the workplace.

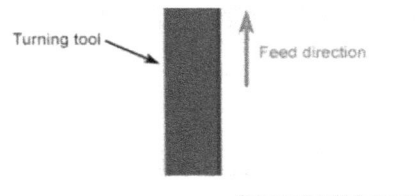

Grooving is a turning operation that creates a narrow cut, a "groove" in the workpiece. The size of the cut depends on the width of a cutting tool. Multiple tool passes are necessary to machine wider grooves. There are two types of grooving operations, external and face grooving. In external grooving, a tool moves radially into the side of the workpiece and removes the material along the cutting direction. In face grooving, the tool machines groove in the face of the workpiece.

Sizing of the incision. The viewport of the cutting instrument. Tool bosses or the necessity to machine are multiple. Grooves that are relatively wide. There are two distinct categories of grooving operations: external and phase grooving. It readily penetrates the soil of the workpiece during external grooving and extracts the material in the direction of the incision. In phase grooving, the tool machines group grooves in the phase of the workpiece. Porting and berthing are machining operations that result in the cut of at the conclusion of the machining cycle. The procedure involves the use of a tool with a base that allows a specific shape to enter, thereby positioning the workpiece perpendicular to the rotating axis. The tool then makes a progressive incision as the workpiece rotates. The tool may be referred to as a "catcher" in order to capture the margins once it reaches the center of the workpiece zone after cutting through the edge.

THREADING AND KNURLING

3D. 3D. These are the exact images that are being transformed and will be moving along the side. What is the subject? The 3D will be sliced on the outer surface. A uniform helical groove of his enormous was filed lints and bitch superior threads require multiple bosses of tool. 3D is a uniform groove. Dissemination. Direction. Feet and trimming. Instrument. Neuralink.

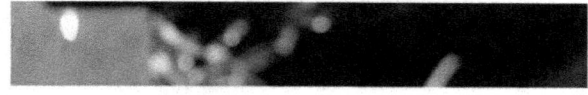

6. Knurling

Knurling operation produces serrated patterns on the surface of a part. Knurling increases the gripping friction and the visual outlook of the machined part. This machining process utilizes a unique tool that consists of a single or multiple cylindrical wheels (knurls) which can rotate inside the tool holders. The knurls contain teeth that are rolled against the surface of the workpiece to form serrated patterns. The most common knurling pastern is a diamond pattern.

The Neuralink operation generates serrated butter on the surface of both noodling and the machine to enhance the visual contour and grieving friction. This machining procedure employs a distinctive instrument that consists of either a single or multiple cylindrical wheels. The process of learning neural networks is. Which can be executed within the tool's grip. Consequently, neurons are composed of. The laborers' teeth are serrated, and they are regulated against the surface. Butter. The most frequently encountered knurling. Subsequently, convert to diamond butter.

DRILLING AND REAMING AND BORING AND TAPPING

During the drilling process. Operation. Extract the substance from the interior of your device. The outcome of drilling is a cavity that is either smaller than or equal to the dimension of zinc. To employ a drill tool. The drill blades are typically positioned on the leaf tool receptacle or the tailstock. Reaming. Reaming is a scaling operation that enlarges the opening in the workings. During reaming operations, the reamer enters the twist axially through the into X, but it also expands an existing hole to the remaining portion of the tool. Reaming is frequently employed to eliminate a negligible quantity of material.

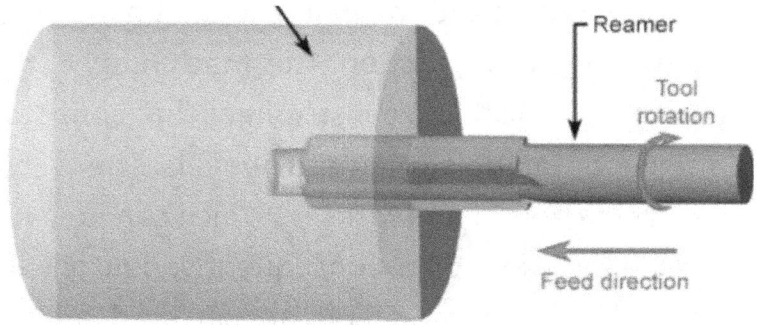

Following the drilling process, a more precise diameter and a finer internal finish are achieved. Boarding. Operation of relocation. It will guarantee axial zorklis and eliminate material from the internal surface to either enlarge an existing cavity or create different shapes or laws. The procedure of TV recording involves the use of taping tool inserts with axial undercuts. The thread is inserted into an existing opening. The opening is compatible with a corresponding bit size that can accommodate. The instrument that is coveted. The operation is also known as taping. If you are unable to, please use the "to me" option.

G00

Okay, let's begin with siblings. Robin. Positioning and traversing G0 zero. The code executes and performs non-cutting movements at a rapid pace. The program is designed to allow the user to specify the coordinate position within the working area or at a specific distance from the previously specified position. Not one. The manufacturer of the machine tool determines the rate of movement. The feed overall control movement on a can be used to reduce the rate of movement from 100% to 0%, but only in increments of 10%. Instructions for executing the G0 zero command. At the optimum input rate, the two slides on the x and z axes operate independently, in conjunction with two full-time buses. In both instances, the G0 zero command will cause both slides to commence movement at the utmost input rate. When the movement of both sliders commences. The instrument will have the appearance of traversing obliquely. A combined movement of both axis. When one axis reaches its finishing coordinate, the other axes will continue to move until it reaches its full completing. As such. This provides us with an impression. The instrument changes orientation, not three.

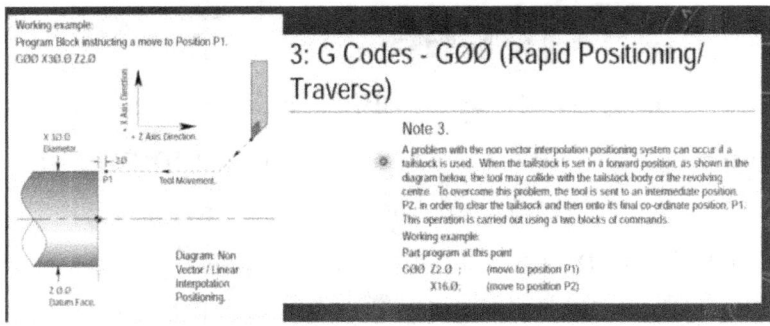

3: G Codes - GØØ (Rapid Positioning/Traverse)

When the stock is set in a forward position, the vector interpolation positioning system can be a keyword if the tailstock is used, as illustrated in the diagram. The following. It may be referred to as the Tailstock body by the instrument. Also known as a revolving center. The instrument is positioned in intermediate position B2 to resolve this issue. In order to resolve the issue and proceed to its ultimate coordination coordinate position, B1. The two units of command are employed to execute the operation. This is the previous example. Note that the tool radius compensation codes G 41 and G 42 should be frozen for the G0 zero. The G001 jazzy is operational when G or 41 or G 42 are active. The tool radius will not resume operation until the A0G01 and G02 and G03 commands are programmed after the AG00 command is executed. The G00 code is modal, modal, and in the force.

Five naught five. Consequently, it is fully compatible with the G01, G02, and G03 codes within the block.

G01

Coogee zero one. Linear interpolation. "G0 one." The code performs a chopping operation. The act of moving. Adhering to a straight line. Additionally, is it three-dimensional? The code's format. In the actual feed, the feed rate value that was programmed into ZG was "0" for one month. Continue reading the post. The axis is iterated over by the toolpath loading. A single axis is moved. The slide will advance at the rate specified. In the G01 command. Additionally, two axes are in motion. The time lines on both transparencies will be identical. In order to minimize the number of vector or diagonal movements. The old vector has been relocated. Separate feed rates will be computed by the machine controller. For both the x and z slides.

Enabling the actual vector v3 to correspond to the state in. The Z01 command allows for the programming of absolute value z axis coordinate movements in accordance with the G01 command. As x and z or incremental value as u and w. Additionally, it is feasible to program one axis with an absolute value and the other with an incremental value. Newton. The following example illustrates the utilization of zero in the command to generate both the diagonal vector and the street. Eliminating the maneuvers. A variety of methods can be employed to program the data in order to achieve this. A functional illustration is provided below. An absolute value and the incremental value for z0. Both g0 one and z are utilized in the aforementioned programs. If the median or predicted value. The Z command allows for the navigation from one item to the next without the need for a visit. The instrument and the object are to be cut. Instrument for cutting. Proceed in a direct line from b1 to b2. Please observe the program feeder. The use of feed override controls can result in a wide range of values, from 0% to 150 and 50%. Please be advised that the machine controller will determine whether 0.01mm is read off if no input rate is programmed on the A01 command. Burbidge on G 99 HMT. Alternatively, if the rate of movement is less than one millimeter per minute on G 98, it is necessary to adjust. Note regarding the barrier. If the block containing the G zero, one, and if he deletes if. Is executed, and the subsequent block comprises, for example, zero and G00. Command. This input value will be stored in the memory of the machine controller. The G00 command is typically used to recline. The input is read to its utmost value. The G00 command will be blocked if it is overridden, and so will any command that contains G01 or G0203. When the original input value is liberated from memory, it will become active. Note five. The G01 is completely uneasy with G00, G02, and G03, as it could model them. Quote within the same block. Tool.

G02 AND G03

From the code 003, we can provide an explanation. Information to be provided. Meaning that is widely understood. What is the direction of rotation? The direction of zero two is required, while zero three is counterclockwise. In the point positions of x and z. In the point position of u and w. A radius or or and input it are the distances from the store to the point. If. Z02 could execute a cutting movement that follows a clockwise circle. The Z03 was capable of executing a cutting movement that followed an anticlockwise circular path at a predetermined feed rate, as indicated by the circular boss departure feed rate. The definitions of clockwise zero, two, and clockwise Z03 are as follows: position it at the coordinates in the diagram below or according to the system. This diagram. Additionally, this is the code G02. Go to the supply store in a clockwise direction. One in the anticlockwise orientation. The terminus of the arc, particularly in terms of the specified equilibrium, is the programming arc. x and Z are identified as absolute positions or you. Additionally, without contemplation. It is an incremental position. Sometimes, this arc endpoint is referred to as the objective position. Utilizing absolute positions x and z. The value is the dimension of the endpoint of the or in relation to the datum position of the component. The coordinate of the endpoint is the distance the tool moves from the store to the position of the or, using incremental position and w. This can be either a positive or negative value, contingent upon the machine's movement. Slides in relation to the starting point. Commence the position. The circular interpolation format for the second program is as follows: Cartesian coordinates are used to calculate the radius of the arc. To permit a clockwise circular circular was absolute and incremental format to follow the anticlockwise circular path. Incremental form and absolute format. G0 two or G03. Defines the direction of the circular interpolation x. Positive or negative.

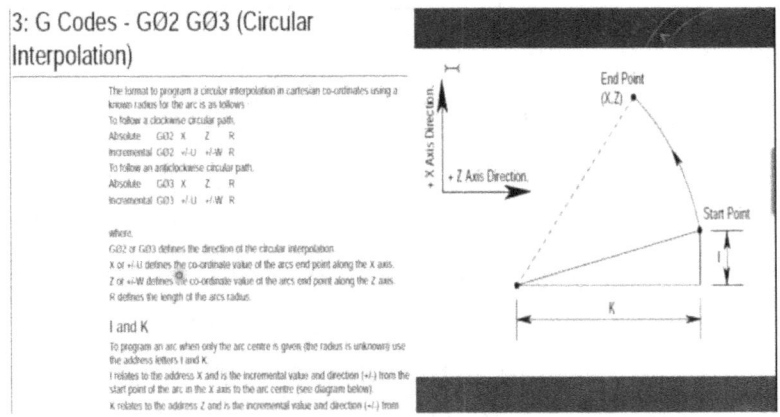

3: G Codes - G02 G03 (Circular Interpolation)

The coordinate value of the or endpoint along the x axis, or for all or defines, is defined by you. The lens of the

Earth radius, or in the event of two programs and an arc, is the owner. The sole information provided is the center of the arc. The radius is either in the key or in the address literals. The incremental value and the other direction are positive or negative from the commencement point of the or in the x axis, two the arc center. This is related to the address x. Key. I am sending this to the address z. And is the direction from the commencement point to the earth in the z-axis user arc, as well as the incremental value. Since. This diagram. Provide an explanation. Inserts. Z02 and the G03 are two examples. As a circular interpolation. This is the format. Additionally, this is the case. Sibling. The transition to the two-three. The distance from the center of the Earth to I is positive. Key. Is constructive. From the beginning to the conclusion. Point. The sibling of the illustration. Circular path movement. Starting with the beginning. To conclude. Everything is favorable. The critical factor is positivity. Additionally, this is the case. One will be relocated by the cutting instrument. Additionally, this. Drawing. Provide an explanation. The. Center of the. Or worth. Therefore, the cutting instrument traverses this. Both. Two in one. Not one. One programming arc that utilizes the read address or the value of or must be equal to or greater than half. The longest distance traversed by either axis. There are no two objects that are greater than 180 degrees in angle.

Two distinct programs or programs that must be implemented. Third observation.

G04

We will commence by providing an explanation of G zero for four d. The zero four code is employed to establish a time delay in the program. This letter x and u for p are used to program the dual value, which is followed by a number that denotes the dual value. Therefore, for instance. The address p is not permitted to contain a single decimal point. The dual is operating at the commencement of the block in which it is programmed. Note three is a dual that commences when the rate of the previous blue reaches zero when the command c is executed. To ensure that the slides remain stationary.

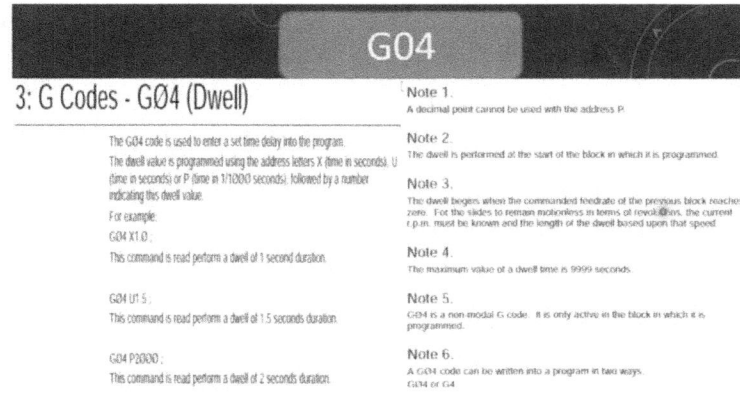

The current or being must be known in the context of revolutions, and the lines of the due will will be established. This implies that the utmost value of a dual is not applicable. Time is. The g zero force is greater than the G designation for nine second note five. It is exclusively operational within the block where its program is located. The final note is that the G zero for code can be written in two ways. The program in two ways: G zero for all, G for.

G20 AND G21

And then I would include six siblings, G20 and G20. There is no difference between one inch and one metric. The standard format allows for the programming of the machine controller in either range, unit input, or metric unit. The program is two for a C in C word, correct? The G20 or G20. For instance, one could be included in the initial two blocks of the program. The unit system of the subsequent item is or is modified.

G20 and G21

3: G Codes - G20 G21 (Inch/Metric Data Input)

The machine controller can be programmed in either Inch unit input (G20) or Metric unit input (G21). The standard format for a CNC part program is to write the G20 or G21 code in the first block of the program.

G code | Type | Units | Lowest input value
G20 | Inch | Inch | 0.0001 inch
G21 | Metric | Millimetre | 0.001 mm

The unit systems of the following items are changed depending on whether G20 or G21 is set.
1) Positioning commands (X and Z)
2) Incremental movement distances.
3) Feedrates commanded by the F code.

Note 1.
The status of G20 or G21 in the machine controller is dependant on the option that is saved to the disc.

Note 2.
A G20 code must not be changed for a G21 code (or vice versa) during the program.

Note 3.
When switching between G20 and G21, the offsets must be set according to the units of measurement being used.

Note 4.
G 20 and G21 are both modal G codes within the same modal group.

The positioning command is established by concluding with the G20 21. Gradual progression. Distance. Feed rate. Directed by the. If it were possible to neutralize, then there would be no one. G20 one is the machine controller, and the machine controller is currently in operation. The. Two codes that are not part of the G20 must be modified when the file is stored. For the G20. One could not only alternate between G20 and G20 during the program, but also between the two. One. The offset must be established in accordance with the units of measurement that will be employed.

G28

And in order to begin elucidating the G20 reference point three, which is regarded as a reference point, iFixit is the machine to which the tool can be applied when machines are installed. This point is also used as a home position. The machine utilizes 2.2 to maintain a stationary position for the Z and X slides. The instrument could be instructed by the G20 to autonomously relocate the tools. For instance, a reference point. The tool can halt before containing and continue towards the reference point by using X and Z to designate an intermediate point.

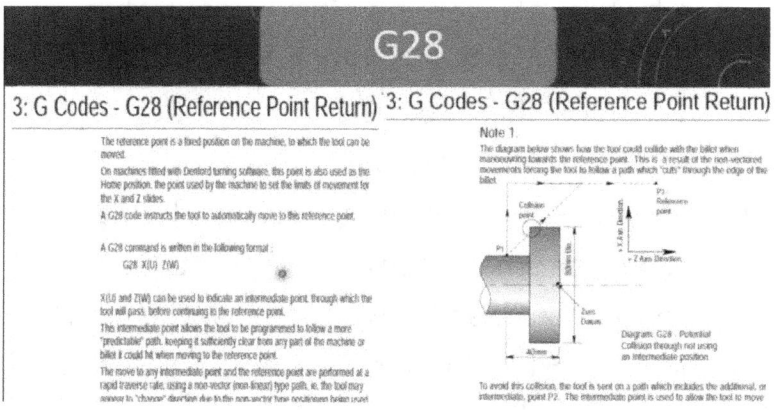

This intermediate point allows for the programming of a new z-tool to follow a more predictable path, while also ensuring that it is sufficiently distinct from any other part of the machine or removed. It has the potential to strike the robot through verse three, or it could strike the robot before it moves to the reference point, intermediate point, and reference point. This vector amusement is well-known. The instrument may appear to undergo a transformation. The instrument may appear direct as a result of the non-vector time decisioning that is employed. Two of the diagrams are located beneath the earth. The manner in which the instrument could chill it with the capacity to move. You are in the process of maneuvering toward the reference point. This is the consequence of the non-vectored movements, which compel the tool two to follow the advanced, which may pass through the pilot's edge. To prevent this collision, the tool is observed on a path that includes the intermediate point B to the intermediate point. This allows the tool to be used completely clear from below before continuing until the reference point, which is three. Schematic spline. Utilizing intermediate points to prevent collisions. The following is an example. Refusing. The instrument is positioned at b1 in the diagram below, ensuring that no collisions are feasible. In this instance, the intermediate entity is unnecessary. Therefore, the following can be expressed. The intermediate point coordinates are still expressed using the incremental address you and the W, but their values have been set to zero to indicate the new x and y axis movement. Consequently, the tool will proceed from b1 to the reference point and then to B3 along the non-vector type path. 3G2 28 is not a model G-code. It is exclusively operational within the program's designated area.

G40 AND G41 AND G42

The G 40 and G 41 and DG 42 fullness radius compensation are a collection of visual and G 42 codes, which enable the machine controller to generate arcs and tables. It is truly axis on the construction. You have the right to automatically provide an accounting for the readership on the tool without sacrificing compensation. The profile could be susceptible to undercutting if it is carved by the instrument. Draw a picture of yourself. The Toulouse projects tool snout radius is illustrated in an enlarged view. Compensation is not necessary in the event that the tool cuts; however, I will apply it to the x and z axes, as illustrated below.

3: G Codes - G40 G41 G42 (Tool Nose Radius Compensation)

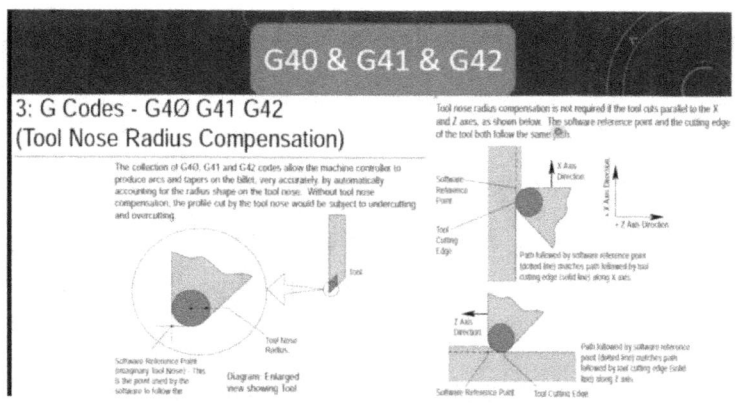

The software reference point and the cutting tip of the instrument are identical. Both. Tables and/or the width of the tool are undercut or overcut as indicated by the direction of the tool's travel in the two diagrams below when it begins to cut diagonally. Instruments of the imagination. The software reference point when the tool is also referred to as the imaginary tools. I am the individual in question. Its primary objective is to streamline the seated procedure. Allow the instrument to remain undisturbed and offset. Point one, the x, and z axes are all included in the pen instrument. Compensation or own portion is implemented. The C in the C can dictate the calculation of the necessary displacement between the imaginary tools and the actual cutting edge of the tool. This results in the instrument adhering to the contour. To ensure more precise programming, both individuals are seated on the platforms, and the distance between them will fluctuate based on the angles and direction. Tool movement is also determined by the radius of the tool point, as demonstrated above. The labeling on the original tool delivery journals can be readily consulted to determine the appropriate nose radius for the tool tip. The following are examples of tool tip code booklets from the ISO standard. The direction numbering in the table is to be lost. The radius is relocated by the tool. Compensation. Directly inputting the tool snout radius and the value of the link. In order to compute the cutter compensation for each tool, the machine controller necessitates three primary pieces of information. This information comprises three components. The volumes to the tool are represented by the values of x and z in the tool offset table number one. Value of offset. That will offset the tool offset value table x and Z, which refers to the tool offset value for the dismantling, transition to, or refers to the tool snout radius. The direction of the tool snout is denoted by D, which is a maximum of 16. The entire offset. The table below contains a list of values that serve as an example of offset. Input from the tool direction number is accessible to designate the orientation of the machinery. The number two, three, and four of tool fasteners correspond to the amount of cutting that is directed toward the shank stone. The knife instrument in the right hand is directed toward the shank stone.

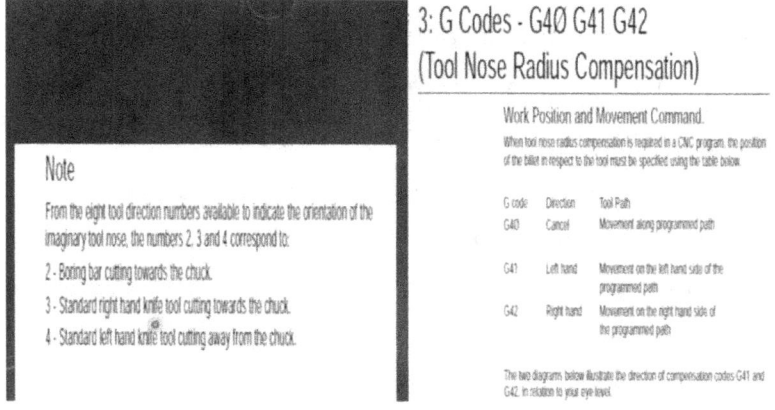

The knife and lift implement that is commonly used for chopping away from. The term "wood musician" references the mechanical point of loss of radius. Compensation is necessary, and the program is the position of the construct, so it must be specified in relation to the tool using the table below. The directions are illustrated in the two diagrams below. The compensation codes are G 41 and G 42. In relation to the level of your lubricant. The direction of compensation G 41 and G 42 is illustrated in the two diagrams below. Regarding your oiling. The movement command and the labor musician. The direction of compensation for G 41 and G 42 is indicated by the four diagrams below. In the second working solution, which. Instrument. Radius remuneration. Commencing the operation. Instruction. Providing guidance. The switch is enriched to facilitate the transition between the prior compensation mode. Is the school packed to block or ramp when the loop is used to enable the entire time to change? To modify. The program is progressing slowly. In accordance with the program's supervision, this line instrument is both favorable. The commencement of the block must meet the following criteria, such as G 42 or 41. It is imperative. 01X and z move forward in the block and the distance of the linear movement is represented by the thin blue or vacancy in the previous block G. The tool snout radius must be greater than the linear motion. The radius value of the tool that is either lost or inserted into the table must remain zero. The orientation number d must be input in the appropriate tool direction. In the table of tool offsets. This is not the circular interpolation command 102403. Anti-business. The Apollo block of the progenitors is devoid of content. Not to include in the utility. Radius is forfeited. Compensation. Begin with two blocks or transition to the C. Control is implemented in C. The initial block is executed, followed by the second block in the third and final hill. In the succeeding composition, the C control has the block, as it moves to block or delete in advance. Performance is currently underway, with the subsequent two blocks stored in memory. This is transformed into this section. This is divided into two.

This is due to the fact that radius compensation is always required for two movements. In the following. This zone is presently being executed. The C in the C control can be converted to a strike in order to determine the correct position for the correct move. That is also the precise position, which will enable the cutter to compensate for the subsequent motion. I apologize; the code G a 40, G 41, and G 42, or Roden, are all part of the same model family Z or are in complete duplicate. The cancellation of two nodes' radius compensation g for g 40. If your foreman is in a block that has a linear move programed, the G 40 that is used to cancel the two nodes radius compensation eg 40 command can only be that. Subsequent to the radius of 44. For central encoders. It is advisable to relocate in exchange for compensation. The next block post must be made within three movements of the previous one, for example, 40 cancellations that can be programmed. Three is the number, and I am only marginally willing to exceed it, which means that the C sequence rule has to be reduced in order to incur compensation. When the machine is initially powered on, it will automatically enter the consumption mode. The reboot button on the controller interface has been enhanced. The program is compelled to execute the boy's performance in the vicinity of zero two, approximately zero, or M 16. The recommended technique for the slaughter and cancellation is the complete robot program for finishing first shown.

G50

And now, we will elucidate the G50 multiple common function claim and the being max maximum rotations formulation. The machine controller interprets the G 50 code in three distinct methods. In accordance with the subsequent language, g is fifty. This command or constraining maximums is used when working with coordinate system configuration and coordinate system shift. We will discuss the imposition of a maximum amount with respect. The figure that follows g 50 is essentially specified. The maximum. The pace of constant surface speed control controllers has been measured.

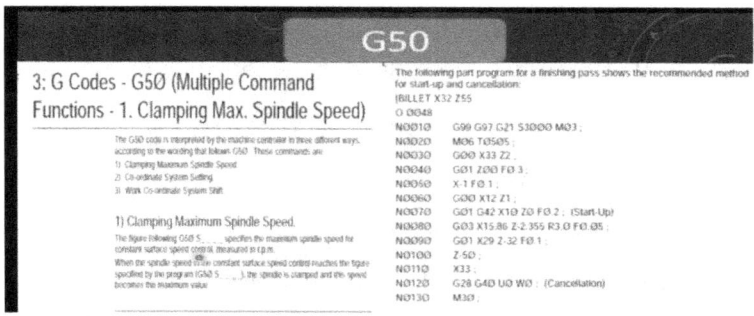

The spindle speed is owned by RBM, and the program specifies the constant surfaces with control that attain the figure. This speed is the utmost value, as the binding is constrained. Therefore, adhere to the program's path to completion. The recommended technique for a top and condition is demonstrated by the boss. Two coordinate systems are implemented. The G 50 code can be employed to relocate the x and z axis along an organization to a different location, as if z were a fictitious location. Graph paper. This is accomplished by specifying the coordinate value that you wish to automate. Supervisor. The tool's current position is to be registered by the controller. This is a tool that will then be a specific coordinate. Despite the fact that it has not actually moved a record, it is a wonderful musician on the imaginary graph paper for the x and z axes. This implies that the instrument itself is not involved. The graph is your portfolio, and in diagram one, zero sits and the face of the billet is aligned with its centerline. Please be advised that the zero digital may be configured in a different location on your machine. The instrument is presently situated at the z 40 and x 30 positions. The zero is located at position zero zero. The number three will be positioned in the earth. The z ten and the x ten should be read. Nevertheless, the instrument remains at the same distance from the billet as it was in the diagram. One. Please review the general G50 command layout for the purpose of setting the coordinate coordinate system using absolute values. The x and z values represent the requisite tool positioning position coordinates. This command is executed. The x and z values should be used to set the present tool position. An absolute perspective on the future. The entire perspective. The diagram tool command is programmed to insert all future absolute dimensions or no relative to a new zero datum position in the instruct and instruct the entire coordinate grid to adjust so that the tool position will align with coordination coordinates. The item and the most straightforward item. Recognize that the command provides instructions. The grid that facilitates the tool's movement. In this command, the entire coordinate grid is moved by reading Z 50 Xin z. Consequently, the position z ten and the extent are now at the present position of the tool. Change in the coordination system. The incremental value can also be employed to adjust the coordinate axes grid, as opposed to absolute values, using the g 50 code. The general g 50 command configuration for the establishment of a coordinate system using incremental values. The command is executed. The coordinate grid should be instructed to move in order to adjust the tool's position. X may be either positive or negative.

G71

removal of one stroke in the x-axis Z of the sibling Z71. ZG 71. In a series of repetitive cycles. Sometimes known as a can cycle. The programming process can be simplified by cycling or utilizing those in the C in C program. This is because only the dimensions that scribing the required component profile are required, and the C and C control will generate the roughing and cut necessary to create the component. Component profile from its internal memory. The movement of G 71 can be illustrated in the following, simplified diagram. Alternatively, or pertains to. If referred to, Robin traverse input. If referred to as "fit U." Depth of incision in the x-axis u to the completion of the allowance in the x-axis w. Finishing allowance in the z-axis. Is the g 71 command routine in the following format? When you. The number of the first block of the program that finishes the shape in the F sequence number of the lost block of the program is the number you wish to reduce in the x axis or scheme skiving amount in this sequence. Completion. Complete the task. Please refer to the x-axis for the distance and direction of the finishing allowance, and the z-axis for the distance and direction of the finishing alone. If you are referring to the feed rate for roughing.

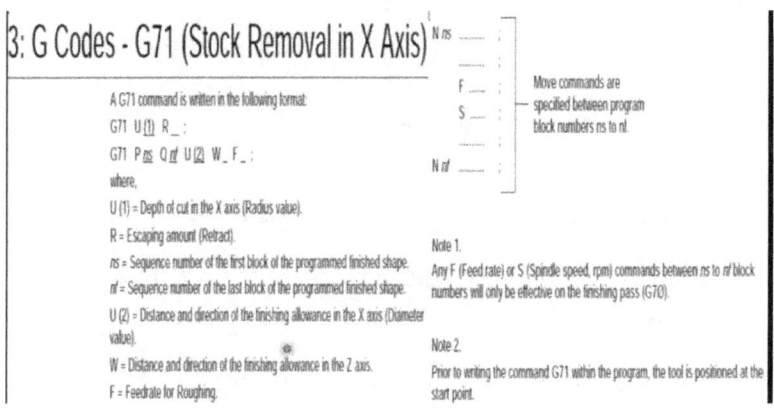

The numbers will only be effective on the final position of 70, as there is no commonality between any group in F. Before providing the command g 71, refer to the broiler. The instrument is situated at the program's inception. In a previous diagram, the sequence of tools in the Taiwan cycle is as follows: is is. Adhere. The instrument retracts away from the start to bring as it moves from position to position. Moves are the distance. It is entirely up to you. Along the x-axis and W along the z-axis. The rabbit feeder must be adjusted to advance the distance you specify along the x-axis. Will. It is centered. Additionally, if he read the response to the z code in the. In this book, the tool moves along the z axis toward the spindle to a point z c in the second tool, which has been calculated in relation to the final profile, the feed read use. The purpose is specified. In the second block of G 7104, the tool retracts in the x and z axes at an angle of 45 degrees from zero, as indicated in the first block of G 71, a Robin Tron traverse feeder.

3: G Codes - G71 (Stock Removal in X Axis)

In the preceeding diagram, the sequence of tool moves in the G71 cycle is as follows:

Move 1 - From the start position, the tool retracts away from the start point. The distance it moves equates to U2 along the X axis and W along the Z axis, at a rapid feed rate.

Move 2 - The tool moves the distance U1, along the X axis towards the billet centreline, at a feedrate corresponding to the G code in the ns block (G01 or G00).

Move 3 - The tool moves along the Z axis towards the spindle, to a point that the CNC control has calculated in relation to the finished profile. The feedrate used is stated in the second block of G71.

Move 4 - At this point the tool retracts in the X and Z axes away from the work, at an angle of 45 degrees. The escaping amount (ie, the retracting distance), R, is stated in the first block of the G71. A rapid traverse feedrate is used.

Move 5 - The tool then retracts along the Z axis, at a rapid feedrate until it reaches the start point of that same diameter.

Move 6 - From this point, move numbers 2, 3, 4 and 5 are repeated by the CNC control unit, until the programmed shape between ns and nf has been roughed.

Move 7 - After all the roughing passes are completed (ie, a rough component shape is visible), the CNC control will perform a single roughing profile pass at the G71 stated feedrate. When this single pass is complete the tool will retract to the original starting position of the G71. The next block will then be read into the CNC control.

Move 8 - If the next block contains G70 P ns Q nf, the CNC control will use the same tool to perform a finishing pass at a feedrate and spindle speed contained between ns and nf.

Should a different tool be required to perform the finishing pass, the block after nf would instruct the machine to move to a tool change position. The next block would instruct the machine to change to the tool required for the finishing pass. The next block would move the tool back to the original G71 starting position. At this point, the finishing pass block (G70 P ns Q nf) would be read into the CNC control. When the finishing pass ends, the tool will retract to the original G71 start position.

It is implemented in these five instances. At a Robin feed, the tool Z retracts along the Z axis. Continue reading until the third point, at which point two objects of the same length will be within reach. The C in C control unit may repeat the second, third, fourth, and fifth moves until the program g between in and then if has been implemented. Upon the completion of all verses, in seven. The year of the component shape is not visible. The C in C will transmit its signal in a single roughing profile bus, as indicated by the G 71. Reliability. The tool will contain the C when the solitary bus is complete. The tool will retract to the approximate beginning position of the C, G, and 71. The subsequent block will be read into the C in C control, as it is the next to contain G 70 B or B or cube. B in this or cu in this. This will contain the between in this and in if, and the c in C control will use the same tool to perform in the finishing bus at the input rate. Is it necessary to employ a distinct instrument to execute the task? The machine would be instructed to move two by the finishing that encompasses the block in F. The machine will be instructed to alter the position of the tool required for the finishing first, and then it will be blocked to return the tool to its original G 71 starting position at the volume. The initial block would be incorporated into the C control at this juncture. The tool will retract to the original G 71 start when the final pass is completed. Musician. Note that G 71 can only be cycled in x x. This allows for the programming of moves between the first two blocks of the finished profile, in addition to G00, G, G01, G0203, and G04. The turns can be utilized to assist with G 71's one outside turn. The tool that is moving toward the outside has been turning two times, while the tool that is moving away from the outside has been turning three times. This has been the direction of the internal turning instrument. The internal turning between moving away from has been this for severing the button or it is treated. The cutting buffer permits the use of either a parallel tool between a and B or a gradually increasing decreasing tool with note one. Diagram. Please provide an explanation for G 71. This is the signal of W and you. G exhibit G 71. An external profile can be programmed using this method. Cycles G 70 and 71. Another example of programming an internal profile using the multiple repetitive cycle G 70 and 71.

G72

About 72, the cycle is comparable to 71, with one exception. The incision is implemented in the z-direction. The diagram below illustrates the complete movement of g 72. Commence at the neutral line. Proceed to the left. Proceed. with a 45-degree upward angle. Continuing in this manner. We are eligible to represent if we represent fit or traverse with the W to complete the final cut in the Z axis. We will complete the x axis cut. The G 72 command is expressed in the following manner. Where the incision of w1 intersects the z-axis. Or, omitting the specified quantity. In this sequence, a number of the first two blocks of the programed completed chip are presented in sequence. The direction of the completion allowance in the x axis and the number of the lost loop of the programed finished chip U distance.

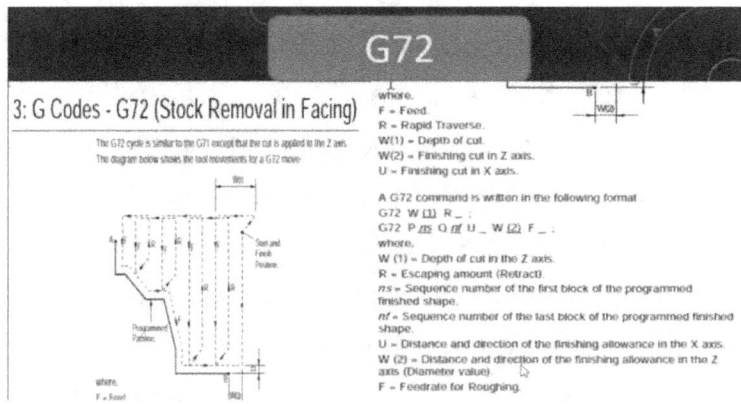

the direction of the finishing allowance and the W2 distance into the z-axis. and if you refer to the roughing, this took about one. The z-axis is the sole permitted axis. Within this section. The key 70 B in is Q is a record that will be read in the C in control after the G 72 of turns 72 command has been completed. This is a note to the completion of us. This method is comparable to the finishing bus one examination command for cutting pattern, which can be implemented in G 72 internal phasing with the z axis cut moving toward the spindle. Subsequently, the z-axis chronometer deviates from the velocity. The external phasing was the z axis cut traveling away from Z, which accelerated the cutting of the material for an illustrated. Beach. Solve the problem in the series.

G73 AND G74

Continue to utilize the three buttons to repeatedly execute the g function. I can cut the program profile through repeatedly, as 73 can. It is primarily employed to machine the board after the initial cut has been formed through crude machining, foraging, or casting. If you are referred to as "fit" or "repeat," please specify the direction of relief and the distance. The format for the three commands is as follows: the x axis, the direction of relief in the z direction, and the w distance. In this sequence, "you" refers to the valleys of cut in the x axis, "w" to the gentle cut in the z axis, or the skimming amount.

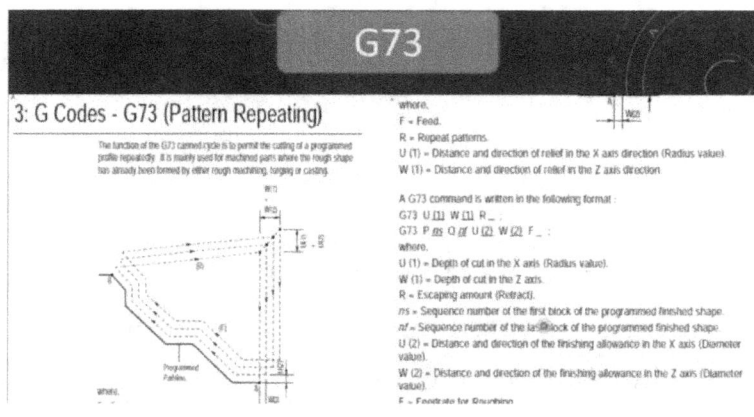

3: G Codes - G73 (Pattern Repeating)

The number of the initial block of the program-finished chip in this sequence. Number of livestock that were lost due to the program's block. The two distances under the direction of the finished over the finishing allowance in the x axis and the distance under the direction of the finishing allowance in the z axis are as follows. If you are nourishing. Not one. the x and z axis movements must be programmed in the. In this block, move to the left and forward, thereby cutting the vector. The void g 71 can also be utilized by G 73. For instance, the programming of an external profile utilizing multiple primitive cycles G 70 and 71. Until its case study. Regarding G 73. G 74 G codes are used in the drainage cycle and phase. The G 73 was only able to perform a feeble training cycle by stroking the machine. This is achieved by drilling the drill with an MC hole in the face of the constructed beak, with the center line of the drill running parallel to the z-axis. The process of removing material is referred to as drilling. The drilling objective is to penetrate the materials and enable the Swiss to swirl, swirl, and clear the rope out before cutting further into the material. Two distinct drilling techniques could be implemented in G 74. The first involves the use of a drill point to bore into a billet, followed by a subsequent movement along the x-axis and further repetition of the operation.

In the event that you require. It is capable of drilling holes that are deviated from the center. Usual. If the wall is the x, b, or permitted from G 74, the drill may be employed. By heating in the stage, a cavity will be produced. Two is the overall profundity. The. The movement of G 74 is illustrated in the diagram below, which is used in conjunction with the drilling of shim blocks. The G 74 command is written in the following format to limit the size of the source, which is beneficial due to the fact that B cycles are utilized. All of us are the criterion element of the bore, where x is the dimension. If step over is implemented. The profundity of this application of the term. Measured in microseconds, be seven units above the x-axis. Q be peaking declines in the z axis measured in microns, and this is for direct unit one G 74. Is she breaking or cycling? She interrupts the cycle. No, if you retract the drilling, I will retract it. Unambiguous. When it reaches the G apogee, maintain the entire value, not the second. The incremental value q may be used in lieu of the absolute values of x and z in AG 74. You and you. Additionally. Example of programming interface drilling or G 74. This is the actual position in the aforementioned program. Is it two at x three? This is the point at which the G 74 command is read into the C in C control control. The drill tentatively advances ten millimeters and then retracts 0.5 millimeters. The complete descent is achieved by repeating these movements three times. At this juncture, the drill advances in a downward trajectory along the x-axis. The centerline 0.0 five millimeter z three retracts and zero b turn traverse have been implemented. Read to the third position on the z-axis. The sequence of b drilling steps is repeated, and the drill travels in a positive direction along the x axis for four millimeters from this point. Enriching occurs when the x-axis enters and the z-axis dips to a halt in 3D. The stop sign is approaching. The utility will execute the action again. The operation will involve drilling each section of the incision until the appropriate support is established. The drill is retracted to the stored position in the z-axis and the cut in portion in the x-axis are lost after all the movements have been completed. The drill is positioned at a specific point in the x and z axes in a move program, and the z 74 is input into the subsequent tool. The drill advances by one millimeter, retracts, and then advances again. Male. Each bond measures five millimeters. All of these movements are repeated until the z-axis falls. Then, rehearse. Drill. Z returned to the orbit. The transverse read is transferred to the starting position, and the next block is read in two ZC in the second cycle. Please be advised that the q value does not contain two. The quality of the total z declines is divided.

G75 AND G76

We will now proceed to elucidate the outer diameter and internal to drilling and grooving of the G 75 code. Therefore, the G 75 common bedroom is a simple cycle that includes drilling and grooving in the x axis, with the potential for excrement breaking in the same manner as drilling at the beginning of the process. There are two capacity constraints. Please provide a detailed explanation of the tool movement in G 75. Also, this is a format for the cold note 20 when G 75 is used for grooving. Select these at the base of the groove. This may be impossible unless a clearance quantity has been provided. The value must be either zero or omitted from the iteration if no clearance amount remains. Resolving an issue. The G 76 is a multiple for reading, and one table uses three cuts to divide it. Two units are contained within Command D. Standard three from the form and which G 74 G 76 used one H cutting to reduce the burden on the tool are all the information required to generate. that is the format for G 76. From 1 to 99, the number of three finished finishing compartments is B. The quantity is depicted in B b. This is the case.

This is the angle at which the chamfer quantity is left by this instrument. The angle at which the tool exits the billet and the end of the three-cutting circle are determined by the angles. The address B simultaneously thwarts tool tips A, b, and C. The number of layoffs is currently resting. Displayed for a quantity of 60 and a tool angle of 60. CU represents the cutting depth in microns when the CNC control calculates the cut depth to be less than the limit. The cutting limits are asserted. Additionally, the utmost value. The finishing allowance is the ultimate or final reduction that is applied to the third. The number of phases required to perform this finishing. The value of B is the allowance for males who are determined. Each finishing allowance is equivalent to the value of the value of or divided by the B e number of finishing compartments. Stage. Ultimately, the position or 3D z of the solid in the z axis b is determined by X z.

Certainly. In this instance, the converse is true. The radius in microns, as measured in the metric system. Q the radius value of the first two buses' depths. What is the circumference of the ring? A minus or defined as the following u and w plus or minus, in addition to not one owing in incremental dimensions or user signs. Determined boys are the channels toolpath or plus or minus the to the money to the wounded boys, the direction of the tool box b plus what was plus Q always plus Look to. For the purpose of systematic. For the sake of symmetry. The movement load city of certain symmetrical buttons may be coincident, contingent upon the sign of the x and z axis. Z 76 common can be used to remove internal threads.

Note 1. When incremental dimensions are used, their signs U, I, J, and W are defined as follows: U and W: Plus/Minus determined by the direction of the tool path. I and J: Plus/Minus determined by the direction of the tool path. P: Plus (always) Q: Plus (always)

Note 2. Four symmetrical patterns can be constituted depending on the sign (plus or minus) of the I and J movements.

Note 3. It is possible to cut external threads with the G76 command.

Note 4. Thread cutting is repeated along the same tool path from rough cutting through to the final finishing cut, so the spindle speed remains constant. The G76 command for compound cutting is active. The surface speed must be used when the cutting cycle is utilized, otherwise the pitch of the thread could be incorrectly machined.

Note 5. When possible, allow a S-run start at the start of the thread cut, to allow for any machine lag run system, etc. Without a sufficient run-in, the start of the thread could be machined with an incorrect pitch.

Note 6. The feedrate override on the CNC control panel will be ineffective, so it is set at a fixed value of 100% during the entire cutting cycle.

Note 7. Although the spindle override feature is not disabled while a thread is being cut, it should not be activated since an incorrect pitch will be generated.

Spindle speed must remain constant, as the node for thread cutting is repeated along the same tool path from preliminary cutting to the final finishing cutting. Maintain a minimal profile. Required to persist. Invariable. The G7 z. If the 3D cutting cycle is in operation, the g 96 command for constant surface speed must not be employed. Alternatively, Wall Street could be at the seashore. Five nodes were incorrectly machined; however, whenever feasible, it is recommended to allow for a five-millimeter run at the beginning of the speed bus. Any lag in the machine propulsion system that does not result in any distress. Insufficient boron. The mechanism was in operation at the commencement of this project. Inaccurate. Which? The feed read override on C in the control panel will be implemented. Infected with. An effective. It is. Seat and the fixed value of 100% throughout the entire thread cutting cycle. Loop. Please be advised that this is a cyclic halt. He will not experience overheating during the trimming process. The thread severing operation can only be halted by using either the reset teeth on the C controller, the button, or both. Not nine. The machine can be configured in this location. It is operated by pressing one block at a time. Pressing the solitary flow. The combustion process is regulated by the T on the C in C. In the event of a single-blow combustion. On completion, one of the D will be operational. The act of reading was. If there is a signal. If. Unmarried. The tool will halt at the being when the back in single loop is activated during the threading operation. Presently, we are at the commencement of the subsequent comprehensive reading bus. For instance, on G 76.

G81

Let us provide an explanation of G80. When drilling with a ball, utilize the standard G80 version. Executes a thorough drilling operation. Here is the procedure. Completely retracts from the opening at the base. The operation known as "one bust cycle" is defined as "zone." If the cycle must be repeated, only the values that change must be inputted into it.

3: G Codes - G81 (Deep Hole Drilling Cycle)

G81 (Diameter Designation, Metric Input) (Diameter Designation, Metric Input) Absolute. Incremental.

N0050 N0050 G00 X0.0 Z2.0: G00 X0.0 Z2.0: N0060 N0060 G81 Z-20.0 F0.1: G81 W-20.0 F0.1: N0070 N0070 Z-35.0: W-37.0: N0080 N0080 Z-45.0: W-47.0: N0090 N0090 Z-50.0: W-52.0: N0100 N0100 Z-50.0: W-52.0: N0110 N0110 G00 G00

The command G81 performs a deep drilling operation, where the drill retracts completely out of the hole at each peck. The definition of one of these operations is called a one pass cycle.

If a repetition of the cycle is required, only the values that change need to be entered into the next block, i.e., the Z depth.

In the above program, the drill is positioned at the start

point and G81 Z-20.0 F0.1 is read into the machine controller. The drill then cuts into the billet to the stalled Z position (Z-20.0) and then rapid traverses back to the starting point. At this point, the next block is read into the machine controller (Z-35.0). The drill rapid traverses forward to within 1mm of the previous cut (Z-19.0), where cutting starts for this pass. These moves continue (cut-in, rapid out to start, rapid back to within 1mm), until the last block containing a Z axis block has been completed.

The next block containing a different G code will cancel the G81 command.

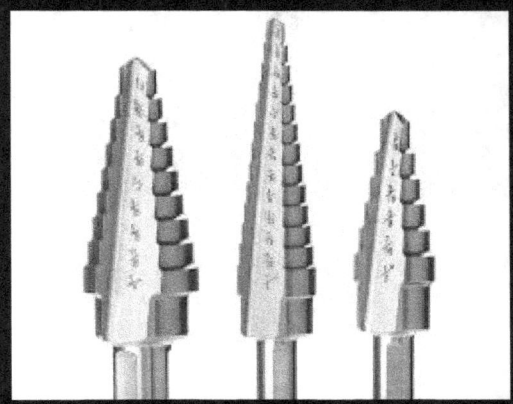

Drill bit cutting a hole

Start and Finish PointZ Axis Direction X Axis Direction 10 15 20 25 20 10 **15 20 25 20 10 15 20 25 20 10 15 20 25 20 10 15 20 25 20 10 15 20 25 20 10 15 20 25 20 10 15 20 25 20 10 15 20 25 20 10 15 20 25 20 10 15 20 25 20 10 15 20 25 20 10 15 20 25 20 10 15 20 25 20 10 15 20 25 20 10**

```
15 20 25 20 10 15 20 25 20 10 15 20 25 20 10 15 20 25 20
10 15 20 25 20 10 15 20 25 20 10 15 20 25 20 10 15 20 25
20 10 15 20 25 20 10 15 20 25 20 10 15 20 25 20 10 15 20
25 20 10 15 20 25 20 10 15 20 25 20 10 15 20 25 20 10 15
20 25 20 10 15 20 25 20 10 15 20 25 20 10 15 20 25 20 10
15 20 25 20 10 15 20 25 20 10 15 20 25 20 10 15 20 25 20
10 15 20 25 20 10 15 20 25 20 10 15 20 25 20 10 15 20 25
20 10 15 20 25 20 10 15 20 25 20 10 15 20 25  10 15 20 25
20 10 15 20 25 20 10 15 20 25 20 10 15 20 25 20 10 15 20
25 20 10 15 20 25 20 10 15 20 25 20 10 15 20 25 20 10 15
20 25 20 10 15 20 25 20 10
```
Sources and related content

Diagram. Represents the mining issue and. If the aforementioned program is executed, the drill is positioned at the start point, and the machine reads 81 z - 20 and feed 0.1. The drill is then controlled, and the built-in two are cut to the specified z. And then, read zero verses. Returning to the original course. At this juncture, the machine controller Z is read in to block. -65. Drilling is conducted from a traverse position. Advance to the point where the cutting stores for this bus are within one millimeter of the previous cycle. This action will persist. Until the final two blocks, which contain zero, have been finished. The subsequent block contains a distinct G-code that cancels the g 81 command.

G90 AND G92

Let us refer to them as siblings. The G 90, which was commonly used, was referred to as the cutting psyche due to its 19 outer diameter and internal diameter. The x-axis is the location where the reduction is implemented. Additionally, during the execution of the command or within the g 90 return block. Tables may be generated. Repetition of a single bus motion is necessary only for the value that must be adjusted in the subsequent block. Additionally, he is. The following is the format for this. For this reason. Not one that pertains to incremental dimension programming. The indications of u and w will be determined by the orientation of the tool movement along both B1 and B2. In the preceding algorithm, there are not two variables: you and W or minus.

3: G Codes - G90 (Outer Diameter/Internal Diameter Cutting Cycle)

G90 A G90 command for straight line cutting is written in the following format:

G90 X(U) _ Z(W)_ F_ :

where, F_ = Feedrate.

This command performs a one pass cutting cycle, where the cut is applied in the X axis. Also, by using the command R within the G90 block, tapers can be generated.

If the one pass move needs to be repeated, only the values that change (in movement dimensioned) need to be entered in the next block.

Note 1. When programming using incremental dimensions (U, W), the direction of the tool movement along paths "P1" and "P2" will define the signs of U and W (plus or minus). In the above program, both U and W are minus.

Note 2. The G90 1 straight cut command can be used for both internal and external cutting operations.

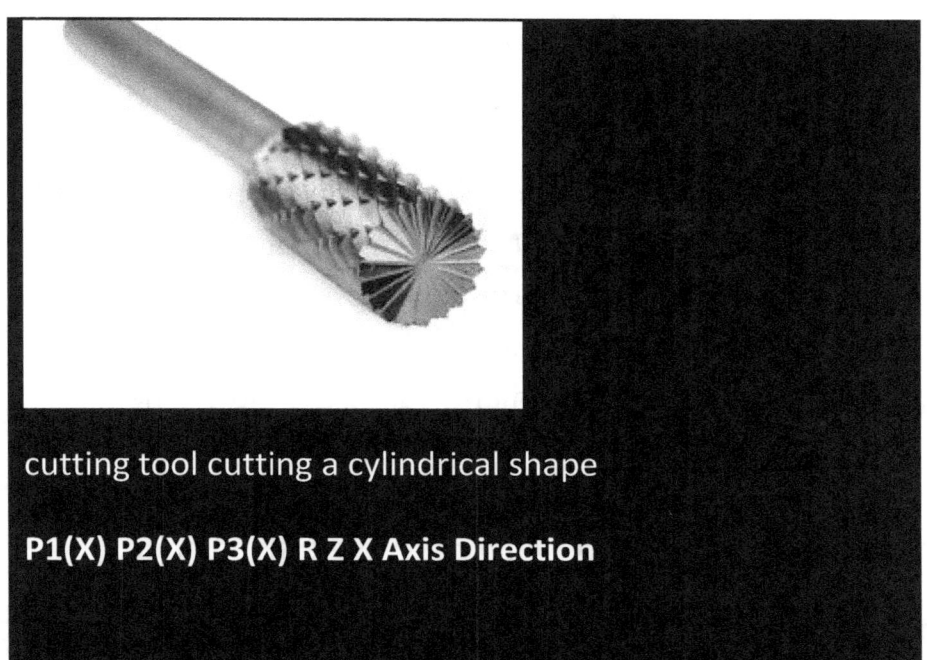

cutting tool cutting a cylindrical shape

P1(X) P2(X) P3(X) R Z X Axis Direction

The G 90. The street cut command is applicable to both internal and external cutting operations. Command 40, G 90. The format of "Were you 40 for cutting?" is as follows. False. The sign for the window on the cutting direction of the bus ins in the aforementioned program, or is in three, is internal as the minute as a minus when the program is not using the incremental dimensions u and w. The direction of the tool movement along path one and the path to will define the sign for you and the W plus or minus. In the preceding program, the characters "you" and "w" or "minus" are used, not "three." The G90 table cut command is capable of performing both internal and external cutting operations. Failure to possess a schematic sibling. When employing G 90. Internal network cutting of G90 outer diameter. Therefore. An illustration of the G 90 program for external cutting. Additionally, the 906 sibling, G 92 three. Cutting in the manner of the G 92. Upon reaching the form and the one-by-zero shredding cycle. The only element that is relocated is the x x. Requires inclusion in the cycle. In subsequent blocks following the initial read-in of the G 92 command, the G 92 command 433 is executed. The format of the final trimming is as follows. G 19 is the schematic sibling. The G 92 Newton is capable of cutting both internal and external threes. The second node. Blogs. Blank cutting is employed with G 92 to generate three finishes from the first two buses to the final finished finish. Additionally, blunt cutting entails the tool nearing the billet at a 90-degree angle to its surface, rather than. Approaching the billet at an angle. In essence, this implies that there are generally no additional stressors. Due to the fact that both edges will be cutting as the instrument penetrates deeper into the material, a greater amount of surface area will be in contact with it. These are our recommendations. The third note is identical to the third note of the three-cut in G 76. The speed run in feet, whether it be all or just the one that is truly relevant at G 90.

G94 AND G96

Ninety-four. Reversing phase. Common performance of the Psyche ZG 94. Phase interruption of the E1 vehicle. The z-axis is the location where the incision is implemented in psyche. Additionally, by employing the term "walk." If there is a repetition, command tables can be generated in GS 94. Repulsion. Is necessary for the relocation. The subsequent group requires only the input of the value z change. For instance, the full format of 94 command four facing cutting psych is as follows. Not one, but when conducting operations. One element of the cycle store store is required for single block mode, but no two Z 94s can be used for internal and external cutting.

3: G Codes - G94 (End Face Turning Cycle)

G94

The G94 command performs a one pass face cutting cycle, where the cut is applied in the Z axis. Also, by using the word R within a G94 command, tapers can be generated. If a repetition of the move is required, only the values that change need to be entered into the next block.

A G94 command for a face cutting cycle is written in the following format:

G94 X(U) _ Z(W)_ F_:

where, F_ = Feedrate.

Note 1.

When operating in Single Block Mode, each move of the tool requires one press of the [CYCLE START] button.

Note 2.

G94 can be used for internal and external cutting.

cutting tool cutting a cylindrical shape

P1(X) P2(X) P3(X) R Z X Axis Direction

The solo format is used to write the Z 94 command 40 versus cutting cyclic. Not a single program employs incremental position. The window direction is determined by the sign of the fuel of u and w, which is either plus or minus. The bus V1 and V2 directions are determined in a manner that is comparable to the method demonstrated for the command G90. You, W, and/or all of the minus values in the aforementioned example. An example program of this reduction cycle is illustrated with the number 94. Resolved. One example is G 94. G 96. Surface B that remains constant. F. Speed modulation. Surfaces. Rapidity. Is a collection of torment. The address is. This is the point at which the pace is determined to ensure that the surface motion remains constant. This foil value is in relation to the tool's position. The unit employed will be contingent upon whether the machine is operating with metric or invariant measurements. The unit of surface speed, or as illustrated. The following is the standard notation for G 96, which is frequently used for surfaces that have control. Format. One should be aware that the programed surface speed may be exceedingly high during the program's execution when the spindle overdrive is utilized. When the machine is in operation, the spindle override control should be overridden. When the range of adjustment already varies from 9:00 a.m. from half of version two, 120. Versions that are within 100% of the programed speed will not be subject to surface speed control.

Note 1.
The programmed surface speed can be altered during the running of the program, using the spindle override control on the machine operating panel. The range of adjustment varies from 50% to 120%, with 100% being the programmed speed.

Note 2.
When constant surface speed control is used, the work co-ordinate system setting and co-ordinate system shift (ie, the software datum point) must be set so that the centre of rotation meets the Z axis (ie, at X=0.0).

Note 3.
The G50 code defining clamping of maximum spindle speed must be used in conjunction with G96, so as not to overrun the safe maximum spindle speed of the chuck.

Note 4.
The G96 code is modal. It will remain active until G97, M02 or M30 codes are programmed into the same block, or the Emergency Stop or Machine Reset is activated.

The coordinate system shift and coordinate system setting must be configured to ensure that the center of rotation aligns with the c z axis. The G 50 code-defined restrict of maximal spindle speed must be used in conjunction with Z 96, as noted in note three. Be aware of the overrun. The shank's maximum spindle speed as specified by the CIF. The G 96 code is not applicable to this model. It will continue to operate until G 97 is reached in 0 two for M0 or M 30 commodities, or until the emergency stop or machine three rest in is activated in the same block.

G97 AND G98 AND G99

Vintage speed, measured in revolutions per minute, is the term used to describe Koji 97. The commander of the G-97. The spindle is currently being expressed in Z units. Revolutions per minute. To be inputted into the machine controller. All subsequent lesbians or defined in revolutions per minute. Following the initial reading. Of the command g 7097. If all changes have been received and program Z is the only one required, its value must be entered. The format of G 97 Command Force Maintenance V. Control is as follows. Please observe that the first note is included. The program that has been received may be classified as L3. While the initiative is in operation.

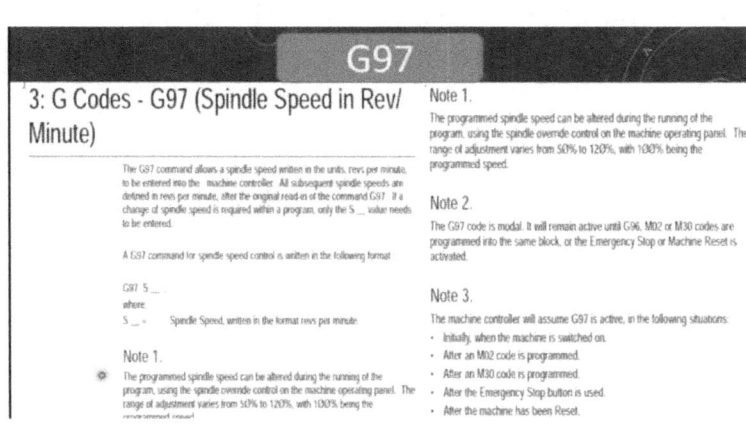

This is equivalent to an override control on the machine. The operation is currently underway. Therefore, the extent of the modifications varies. 50% to 120% with 100% indicates that a program is not being implemented. G 97 is in the process of being relocated. It will remain operational until G 97 M02 or in 0730 code or program in the same block, emergency halt, or machine resetting activity mode three. By 97, the machine controller will assume G. Is engaged in the subsequent scenario. Initially, the emergency stop button is utilized to program M02, which is then programmed in three zeroes after the machine is turned on after n. Following the machine's reset. Let us discuss G 98. The rate of G is 98 beats per minute. The G 98 command. Reverse. Millimeters or inches were intended to be input into the machine control if the unit is deleted. All subsequent feeds that are read or defined in the rules. Within the selected unit.

3: G Codes - G98 (Per Minute Feed)

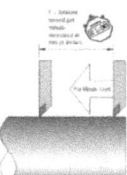

3: G Codes - G98 (Per Minute Feed)

Note 1.
The programmed feedrate can be altered during the running of the program, using the feedrate override control on the machine operating panel. The range of adjustment varies from 0% to 150%, with 100% being the programmed speed.

Note 2.
The G98 code is modal. It remains active until G99, M02, M30 Emergency Stop or Machine Reset is activated.

Note 3.
The machine controller will assume G98 is active, in the following situations:
- Initially, when the machine is switched on.
- After an M02 code is programmed.
- After an M30 code is programmed.
- After the Emergency Stop button is used.
- After the machine has been Reset.

The only original read in the command G 98 is the one that follows the original read. All that is necessary to interpret is the if value if a change in feed rate is necessary within a program. The unit. The. Determine whether the machine is operating in metric or imperial units by utilizing the unit specified in the window. The following format is used to write the G 98 input rate command. Note that the feed rate can be used to supersede the control on the machine operating component, allowing the program to be modified during its execution. The programmed pace is available in a range of adjustment from zero and 0% to 184 and 50% within 100%. Different from the G 98 code is the model. It will continue to be active until G 99 or G zero to M02, or in three zero merging systems. They remain in place or the machine reset function is activated. Third observation. G 90 will be assumed by the machine controller. It is operational in the subsequent scenario. Initially, the machine is programmed with zero code after a M series when it is in the presence of minors. Following the activation of the emergency halt switch. Following the machine's reset. The command is G 99. And now, if it is read in seven millimeters by revolution or inch version in the future, it will be entered into the machine controller or subsequent feed readings or defined in the chosen unit after the original read in of the command g 99. If a change or citrate is necessary solely within a program, the units used to enter a value will be contingent upon the machine's operational status. She was employing either an imperial or metric machine. The G 99 input rate command is expressed in the following manner. One point to consider. The input rate can be adjusted during the program's execution. Control the machine's operation by overriding the control. The adjustment range was 0% to one, with 50% falling within the range of 100%. The pace that has been programmed. Model is the concluding note of the G 99 code. It continued to function until G 9898 M0230. Stop in the event of an emergency. When the mechanism is in operation.

M00 TO M09

We will commence by elucidating the zero zero program halt on the machine controller. Reading the code in zero zero, was there any blue? It is the repository for the program. The program must be contained by pressing the key. M01 is optional. There are two. The M01 has the potential to fulfill the same function as the M00. The machine controller exclusively acknowledges the singular. Establish the signal to maintain the program. If the halt input control is enabled. M02. The program's own code denotes the conclusion of the program and executes the general reset function. The machine controller could also function as M05 when the C in C denotes its initial state. M03. The forward should be spun.

4: M Codes - MØ3 (Spindle Forward)

MØ3 - Spindle Forward (Clockwise).
The clockwise direction of the spindle is determined by viewing from the back of the machine headstock, along the Z axis towards the tailstock.
The spindle start command is activated at the beginning of the block in which it is programmed, ie. before any axis movement occurs.

4: M Codes - MØ4 (Spindle Reverse)

MØ4 - Spindle Reverse (Counter Clockwise).
An MØ4 code acts in the same way as an MØ3 code, only the spindle rotates in the opposite direction.

A ceremony is conducted in the clockwise orientation, but the machine is viewed from the rear. It is ensnared along the z-axis in the direction of the opening that will halt this boundary. The spindle stored command is activated at the commencement of the loop in which it is programmed. Prior to the migration of any X. M04 has been the opposite, and it has the potential to act in the same manner as M03, but it has been rooted in the opposite direction. The rotating in activated at Z in the block in which it is programmed, and M05 has been there since it was unable to cease. Following any axis movement. The m is changed automatically by the M06 utility. The machine is activated by the code. Forward and forward, thought and it. The lads were unable to instruct it to proceed. Stated. Two digits. The format of G in M06M0. The refrigerant could not be changed on this device. explosion, explosion. The code switch in the coolant pump of M09 coolant.

M11 TO M99

Within. Schenck. The use of PowerShell lm 11 will be closed upon opening this code. The power chain view will be closed by this code. In 13 spindles for refrigerant tone. This code integrates the functionality of M03 and M0E. The M05 code will halt the spindle and coolant in the 14 spindle, reverse, and coolant. Owning this code entails the same functionality as M13 16, but the spindle rotates in the opposite direction. Stop coil exits at M 25. It is this code that will operate the stop quote. The spindle must be halted. To permit the machine to enter. Activation of the controller is scheduled for 25. In the year 26. Tailstock. When the retract. The till stop coil will be specifically driven by this code. This implies that it must be halted in order to permit the machine controller to activate, and it must be done in 25 code. I am in the process of inserting the program, stopping, and resetting the M60.

4: M Codes - M25 (Tailstock Quill Extend)

M25 - Tailstock Quill Extend.
This code will drive the tailstock quill out (extend). The spindle has to be stopped (using the M05 code) to allow the machine controller to activate an M25 code.

4: M Codes - M26 (Tailstock Quill Retract)

M26 - Tailstock Quill Retract.
This code will drive the tailstock quill in (retract). The spindle has to be stopped (using the M05 code) to allow the machine controller to activate an M25 code.

This could potentially prevent the program from properly executing its sync signals. The control is subsequently reset to the program's beginning, signaling the end of the program. If the block number is followed by the lm 13 insert code, the program will be reset to the specified block number. For instance. This command. This command is executed. Reset the program to block number 140 and cease its operation. In 40 words, the lm 30 could also function as a M05 and M09M code. Draw or expand. The word sketch of I would is driven by this code, which causes it to open beneath the word barrier and rupture. The collector retracts at 41 baud. This has the potential to return the birdcatcher portrait to its operational state following its explosion. From IL 62 to 66 and the M 77. Auxiliary output function. This could enable the transmission of a signal from the machine controller to the various devices. Diverse devices, including robots. Then, await the return signal, which will indicate that the device has successfully executed its function. In the year 1990. Calls from eight subprograms. This code will cause the machine controller to transition from the main program to access a separate program in its memory, known as a sub program. The code from the 99 sub-program is inserted into the previous sub-program when the missing line of the sub-program is encountered. Restoration of control to the primary program. A continuous sequence will be established if an Im 99 code is programmed at the conclusion of a meal program. Control will be returned to the program line with the same number if an Im 99 code is followed by a block number. As stated in B.

MASTERCAM AND COURSE INTRODUCTION

Squid Cam will be the subject of this project, and we will then proceed to the master game. a few features in their applications for CAD cam. Initially, our goal is to comprehend the necessity of CAD cam infection. For what reason do we do what we do? Are you frightened, cam? the reason we are not adhering to the traditional approach and then. Okay, the subsequent step is to be able to generate 2D geometries in Master Plan. Therefore, the initial subject that we will study in the Master game will be 2D geometries. We will transform these 2D geometries into 3D representations. Additionally, we will generate tool components at the conclusion. Essentially, there are two categories of tool parts: 2D toolpath and 3D tool. However, 3D tools are not presently included in your course. However, we will acquire the ability to generate 2D tool elements in matter in and. Lastly, we will transform these tool elements into an encore of a CNC machine. Okay, so what is CAD cam? Before we define it, it is important to understand the concept of a case. Thus, computer-aided design (CAD) is the computer system that facilitates the development, modification, analysis, or optimization of a design. The term "kit" is primarily employed to refer to the process of creating and modifying a model. Additionally, the term "computer-aided engineering" is employed to optimize and analyze design. The utilization of computer systems to plan, manage, and control the operations of an infection plant through direct or indirect computer interface with plant resources is known as computer-aided manufacturing in the context of games. Therefore, the game is not merely a matter of creating a model, creating a toolbox, and converting the toolbox into entity code. Essentially, all devices in contemporary facilities are connected to a central computer.

CAD/CAM

- Computer-aided design (CAD) is the use of computer systems to assist in the creation, modification, analysis, or optimization of a design.

- Computer-aided manufacturing (CAM) is the use of computer systems to plan, manage, and control the operations of a manufacturing plant through direct or indirect computer interface with plant's resources.

Therefore, upon the creation of any model, process plan, toolpath, or NC code, it is transmitted to the central computer, which subsequently assigns it to a desired machine. The term "cured game" is used to refer to the complete management and control of all processes.

Initially, the 2D model was attempted to be converted into a 3D model, and the art was subsequently manufactured using a computer. Why is it necessary to employ CAD cam, and why are we not continuing with our conventional manufacturing methods? Therefore, gate cam enhances the productivity of the design and the manner in which it does so. Once we have completed the design of a model, we will proceed to analyze it in CAD, which is once again pertinent to CAD cam.

Need for CAD/CAM

- To increase productivity of the designer
- To improve quality of the design
- To create and test toolpaths and optimize them
- To help in production scheduling and MRP models
- To have effective shop floor control

Therefore, we design and analyze a model in a CAD. We do not require the model to be manufactured for mechanical testing, as this reduces the overall development, product development, and analysis time. The second objective is to enhance the design's excellence. Therefore, the majority of CAD software includes a shape optimization feature that enables us to reduce the model's weight and optimize it in a variety of other ways to create, distribute, and optimize it. Therefore, we execute manufacturing operations. Additionally, software. Then, we optimize them to reduce the time required for production scheduling by ordering them and optimizing the cutting parameter. And it may be beneficial to establish a more efficient schedule and effectively manage the facility when the entire arcade game is connected to a central computer. The process of developing a game system. All right. So, initially, we establish a park geometry. Subsequently, we establish a tool database, which is referred to as the "library," which contains all of the tool's information.

How do CAD/CAM systems work?

- Developing NC code requires an understanding of:
    1. Part geometry
    2. Tooling
    3. Process plans
    4. Tolerances
    5. Fixturing
- Most CAD/CAM systems provide access to:
    1. Part geometry
    2. Tooling

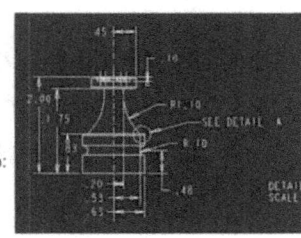

We have the ability to manually enter it into the rules detail; however, the majority of the time, we are provided with this information by the tooling companies. Therefore, it is feasible to generate two- and three-dimensional designs on the master game's mask, while simultaneously translating other files. I simply have 2D drawing files, whereas STL files are 3D files with coordinates. Step is primarily employed for the creation of a 3D solid gate model, whereas STL is employed for the formation of meshes toolpaths. So Master Cam is regarded as one of the most exceptional tools for generations of hair due to its extensive tool library, which encompasses a broad variety of tool, machining, and parameter selection. The master can assist us in optimizing the machining parameters. NC code can be generated with a single click in the C program generator. The most critical aspect of our manufacturing process is the ability to animate it in order to visualize the machining operation instructions for a generic NC machine. Therefore. Once all machining operations have been completed in the master can, an encoder is generated and the code can be edited in the master cam code editor. Lastly, we transmit the code to the machine once it has been finalized. in the process of repairing.

INTERFACE PART 1 SKETCH, MODIFICATION TOOLS AND TRANSFORM

Prior to commencing that, please inform us of his intentions. This is evident from the name. The term "cancer fear" is typically employed in the context of machining. The imaging process may involve lathe milling, plasma, or laser cutting. Consequently, we develop a substantial part and subsequently execute all machining operations in the master cam, generating a code for this quadrant. This implies that the image in the component is synthetic. Therefore, we should commence with the units. Navigate to the Configuration file. And here, the units are in centimeters. These units are intended for analysis. These are units of analysis. Therefore, I will establish a limit. Additionally, configure it to use the metric system. Given that you are aware of its metric. All right. Initially, we will commence with the wireframe, which is essentially a 2D representation of the majority of these operations. Additionally, we will address these 3D operations at a later time. Therefore, we should commence with the discharge point's location. This command enables the creation of points at any location on the screen. The creation of touch locations is visible.

These markers can now be employed for the driving lines. Additionally, we will observe the exercises in numerous other versions. It is acceptable in certain respects. Therefore, the line could be classified into three categories based on the number of elements, which ranges from one to two. Next, we will examine the midpoint line. I will choose the midpoint in this instance. Additionally, the length is symmetrically included on both sides of the designated point. In this manner. Margin is the third sort of line.

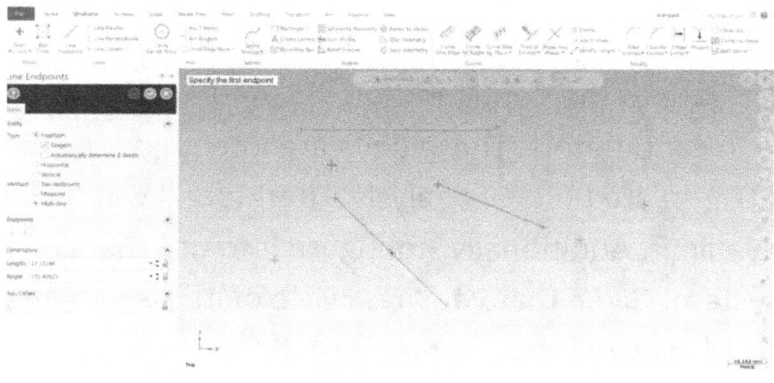

Therefore, the command is not deactivated until the operation is completed after it is activated. In this manner. The command is not concluding lines until we terminate it. All right. It is now time to discuss the second form of point, which is the dynamic line. It is employed to establish a point at any position along the line. Subsequently, I will select this line and determine whether it is possible to select any point. Assume that this point is established in this manner. In this instance, we can establish point eight of the untitled, but we are unable to quantify the distance. If we wish to indicate this precise distance, we will employ a point segment. Therefore, I will now select this line. All right. Let us assume that I wish to partition it into four sections. For instance. The line is now divided into four segments, as is evident. All right. In the same vein, we are uncertain about the number of segments into which we wish to partition the line. Then, will we employ the distance? For instance, suppose I wish to include an additional point following the word "lemon." Therefore, choose this line.

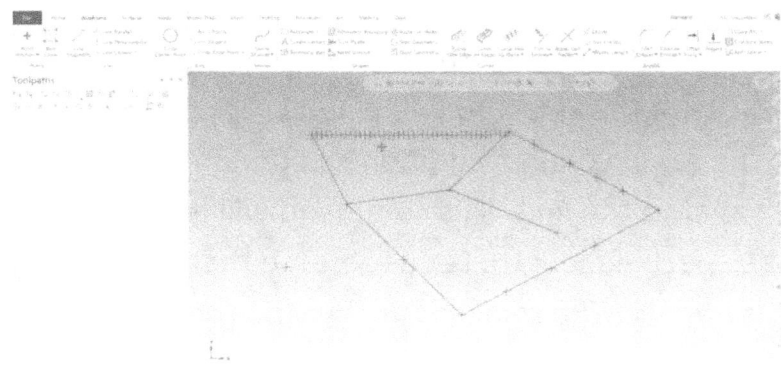

Additionally, there are numerous factors to consider. The third variety of point is the point terminal point. Therefore, we should eliminate all five of these items that encompass the time. I will now draw some lines in the following manner. Additionally, we should activate the orange line. Now, when I select this command, it will accumulate points at the conclusion. The conclusion of each sentence. Observe that points are generated at the conclusion of the line. We can proceed to the subsequent point, which is null. utilizing speed markers. Therefore, we will address it in three paragraphs. Therefore, we will transition to the line types for the time being. First, the parallel line is created using the parallel line command. Lines that are parallel to the extant line can be established. Let us assume that I requested a distance of fifteen meters. All right. Presently, it is situated on this side. However, if we wish to accommodate the. Therefore, I will select the opposite side to observe the line's velocity. Additionally, the bottom line is generated on both sides of the selected line when I click on both sides. In the same vein, an additional command is perpendicular to the line. This command will generate a line of the specified length. Let us assume that the number is twenty. The selected line will be the edifice, which I have chosen.

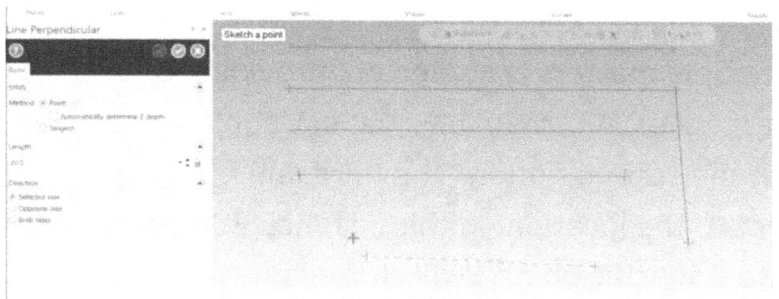

I am now able to specify any location. In the same vein, we can incorporate an additional perspective. On the opposite side and on both sides. All right. The third line is the closest, for which a circle will be required. Initially, it is necessary to establish the definition of a circle. Afterward. I have chosen the centerline. Afterward, we will establish a mean radius of 30. Therefore, I will now establish a boundary. What is it that I desire? I am seeking to establish a connection between this line and this circle with the least amount of space. Therefore, I will choose the line that is nearest. I will choose this limb and this item. See the line that you are referring to. The smallest distances that can be generated. All right. Subsequently, the line is bisected one at second. And this sentence. The line is in a state of gravitation. All right. Increase the number to sixty. And if I desire multiple lines, I now have the option to select either one of them. Let us assume that I wish to retain this line. If I select "single." Therefore, the desired line between the selected line is generated when multiple masters are selected, providing us with four alternatives. Alternatively, we may continue to maintain the second option. The third line is the line that is tangent to the location. In this instance, it is possible to designate a specific point. I will choose any point. So, let's do this. We can now observe the formation of the tangent line. I am now able to access this file. This is where it is saved. I am now in need of an additional line. Therefore, I will consent. Create this operation. The second.

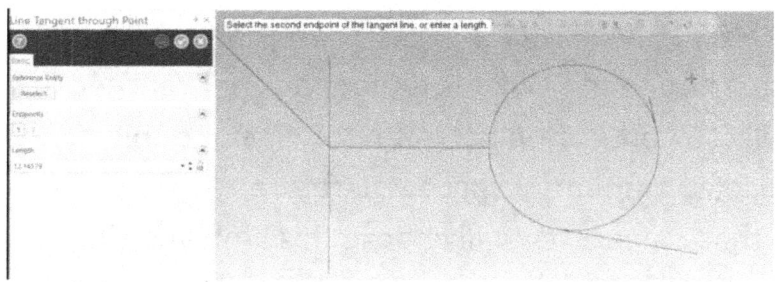

Furthermore, it is intriguing in design. This is the case. Remove these. Next, we will discuss the 2D point. This enables us to generate an arc that utilizes three elements. Therefore, the ship is under the control of the first, second, and third points. Additionally, we have already established select import. All right. Therefore, an additional function that is associated with the is more closely related. Assume that we have an existing system. Are we planning to shutter it? I will end it by clicking on it and making a selection. The arc is a closed-form circle. All right. Currently, we are engaged in a discussion regarding an additional form of name. It is the motor arc. Therefore, I require a few sentences. Certainly. Additionally, is this a vertical line? Is this the engine arc? Initially, I possess this line, and subsequently. I am required to identify a tension point, which I will refer to as this one. Right now, I have two alternatives for selecting an arc. I am capable of selecting any individual. We should retain this item. Therefore, we should accept this. All right. It is acceptable to conduct a circle at this time. In this instance, it is possible to define a circle with three vertices. Let's save this now so that it can be activated. Therefore, I was given a pseudonym and began to print circles. Currently, if I wish to construct a circle, it will be tangent to other bodies. Therefore, in this instance, I will choose one name from the second round of the circle that is tangent to this line. I can now specify the name of the circle to be 15 movements below. If I am able to reach thirty at this time, it will be. Certainly, I will extend the radius. This will decrease by fifteen. You can observe that this is being moved to this location. Now, if I wish to generate three points, let us generate them. This is equivalent to this.

Therefore, it will be straightforward to comprehend. As I, this second. Enumerate. Regarding. Microsoft Word. Therefore, in order to incorporate a circle between three points, I will select three points. I apologize for the three points. Perpendicular. I would now choose this item, this item, and this item. Therefore, a circle will be generated automatically in this scenario. Additionally, we will be unable to modify its designation. We refrained from reducing. All right. So, yes. If I wish to create a circle with three points, I will adhere to this initial point. The second point is now to be addressed. The third element is the control of its position. Additionally, the radius. Therefore, we should arrange it in this manner. Additionally, we neglected to address two critical concerns. Therefore, we should address two points.

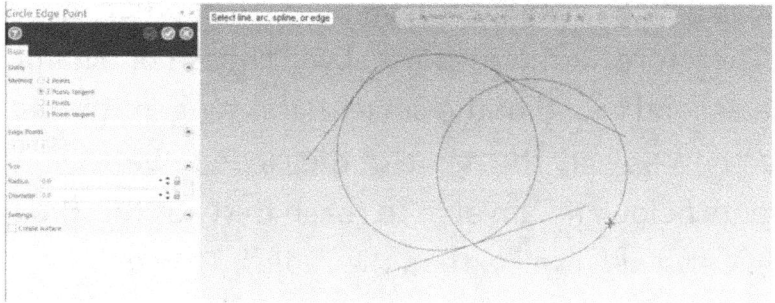

I will initially choose this appellation and subsequently state that the village has been established. I can now modify its location by utilizing its name, as shown below. Novel forms. The subsequent item is innocuous and is employed for the purpose of creation. This is followed by the line, which is employed to establish a consistent shape. Therefore, there are various varieties of lines. The initial step is the minimum. Therefore, I will opt for this. Additionally, I am unable to provide any further details regarding this matter. In this manner. To generate any geometry. All right. Now, I will generate and subsequently comprehend the second advent. This and another type are the blending lines, which are generated by combining two existing lines. Therefore, I will refer to the first curve and any point on this curve as the second. Afterward, a second point will be made. Therefore, it will be mechanically transcribed as the line in accordance with the current state. The following are a few of the positions. Trimming is the default setting. So, if I uncheck it at this time, all of the lines will be present. Now, if I verify the format. Therefore, this is the current trim.

Therefore, voting trees are pruned. If I choose one, it will be one of them. The data is trimmed. The second scenario is when I select "break." Therefore, one is obligated to undertake an inspection, which in this instance pertains to an entity. It is divided into two components. If I select entity one. Therefore, the initial entity is divided into two components. If we both observe, the two are divided into two components. Therefore, we should administer it in this manner. We should address this matter, which is benign to leave to them. Therefore, we developed this benign operation in this manner. I will now proceed to the point of movement. It will generate elements, which are the foundation of the set. The scores will be linked with the assistance of, which will be generated. This phrase. Therefore, the scores are generated. This is not particularly beneficial for identifying the critical points in the line. Therefore, that is required. The subsequent function is the bold circle. So, let's begin with the complete circle. I have designated eight as my number. Therefore, I will choose any base point in the same manner. Additionally, an additional individual is required to establish the screen. Therefore, the outer diameter is 200 in this instance. We have the capacity to adapt. Let us amend the line to 56. All right. Subsequently, the central point is the primary focus. Therefore, it is ten. This diameter will be ten if I reach 15. All right. Therefore, the type is arcs and points, and both types are available when I select points. Therefore, these are the sole locations where points will be generated. View the objective. Additionally, the points will be generated by clicking on both of the circles.

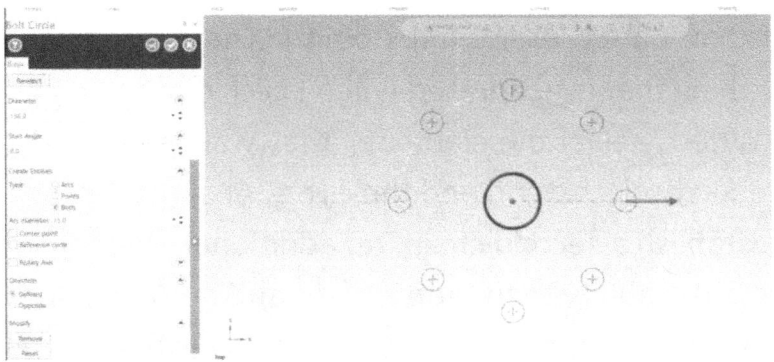

In the same way, it is possible to designate the start and end angles. If I select "Start now," I will commence at 45 degrees. Consequently, the circle will be repositioned. Additionally, a smaller circle serves as an essential function. Therefore, there are certain instances. Initially, we can select the full circle in terms of its number. Therefore, we should assign a name to this entity. And now, modify it to "Let's say I wish to advance." Therefore, there are four cycles. And I can specify the darn thing. Therefore, we should proceed with that. In this manner. The start angle is 60, and the value is 115. Therefore, I can modify it to zero. As you can see, everything is now in its proper place. All right. If we wish to gently rotate it, I will adjust its angle to 45 degrees. All right. Currently, these are being rotated away from the origin. In the same vein, the subsequent item is. Movement. It is now evident that both individuals have made their selections. Both refer to the circles that are present in the points. The circles are the only thing that is generated when I click on the button. When I select the elements, they will be generated in both the grid and the circle. All right. Next, we will establish a second chain. An additional function of the class is the ability to identify angles and numbers. In the initial scenario, it is always possible to specify a numerical value. In this instance, it is possible to specify both of them. Let us assume that I employ three circles at a 15-degree angle.

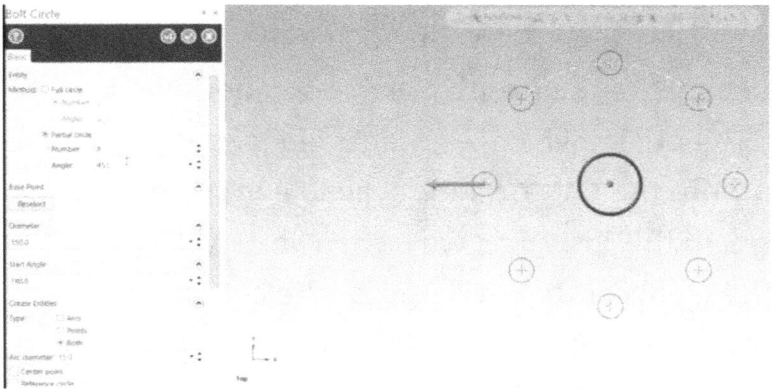

Therefore, this portion is the only one that is utilized if I increase the amount of information, for example, if 30 individuals did not approve of this. If I raise it to ninety. Therefore, half of the space is currently utilized. All right. Another rectangle is currently in 2030. Therefore, rectangles may be of two varieties. One is equivalent to two. By selecting these two opposing points, we can generate a rectangle. We have the ability to regulate the dimension in this location. Additionally, I intend to incorporate an additional operation. However, this time, I intend to generate a rectangle from the center point. Therefore, it is evident that the rectangle is generated from the center. Therefore, we have the ability to exert control once more. Now, what if I desire these corners to be rounded? So, we can implement two methodologies for these. I shall be the first to implement it. Initially, I will execute this command in order to create rectangular shapes. This standard is another method that we will discuss at a later time. All right. So, initially, navigate to the shapes section. Currently, we have a plethora of options. The origin, which was centered, and the opposite columns were the only options available in the previous command. In this instance, we have a variety of alternatives. Therefore, I will opt for the center. Additionally, it is angular, and the midpoint is straightforward and rectangular, as represented by this image. Presently, there are three positions. We must choose two points that are in opposition to one another. Presently, I desire the extremities to be encircled. Consequently, I will adhere to this program. Therefore, it is necessary for me to specify the radius in this instance. We will choose a radius of two. The rectangle that was selected is then added to this grid. Moreover, if I wish to apply this in a single manner, as illustrated in the image and in the box. It is calculated in this location. Now, if I wish to double-click on this and advance to the center point, I will do so. Also, there is this. Therefore, to establish a connection. All right. Now, there are additional shapes, the first of which is the polygon. Therefore, it is six-sided by default. Polygon. The central point and radius are the subsequent components. Currently, it is evident that the polygon is not contained within the circle. If I desire, I can position the polygon within the center.

Therefore, I will select it and observe an additional polygon within the circle. It is the same with the pointed corners. Corner-centered is zero. If I wish to modify it to, say, three, that is acceptable. Three can be observed. Let us increase the number to ten. All right. The comments are now evidently incorrect. Additionally, if I desire these corners to be situated in a different location. Therefore, I may rotate this. For instance, it is currently rotating. The subsequent observation is that we should proceed with this. The subsequent operation is. It is now time to select the central palm. The neck's radius is now visible in the x-axis. This point is being established for me. I am now required to specify and obtain a license. In this manner. The segments and line segments that are generated are seen. These are a sophisticated alternative. Therefore, we will refrain from proceeding to this point and instead modify our parameters. Setting them to sixty is the result. Additionally, this pertains to one. If we desire a portion of this requirement. Therefore, we have the option of establishing a specific starting point. Subsequently, let us assume that 90 is the final number. Therefore, it is equivalent to three times four thousand. In the event that I select one, it will commence. Consequently, the situation will unfold as follows. All right. Additionally, we have the option to rotate it. Let us assume that this is 45. All right. Delete it. Subsequently, we can import it using a 3D card. Therefore, we have not addressed this matter in this context. By that time, the subsequent shape will have been established. Therefore, the interior radius of this circle is denoted as radius. I am the one who is from this point to this point. Revolutions are the number of actions in one, two, three, four, and five. All right. The vertical distance between these two lines. Initially, it is ten. Ultimately, a distance. Therefore, the distance between each of them remains constant. Initially, I would like to be five, as in the final. Therefore, we are unable to observe this. One. False. If you are able to observe the transformation of. For this, I must establish it as a single entity. The initial value, which is one, is now visible, and it is gradually increasing to ten. In the same way, we can apply it to ten and one. Additionally, there are ten M1s. The current value is one at the outer edge and ten at this location. In the same way, we can alter the direction by experiencing varying degrees of discomfort. Again, clockwise attention is directed toward clockwise. All right. Let us examine some modification operations that incorporate a slight chamfer in the boxing process. Therefore, we should create a rectangle. okay. Additionally, tap one to generate it. Subsequently, if I wish to execute the pattern, we are unable to employ the rectangular shape command; rather, we must employ a square with an incorrect corner. Therefore, we will implement this command. Therefore, I will choose these two commas. Presently, it is 510. Let us achieve it.

Presently, this trim is being established. This line is now visible if I uncheck this box. Still, it remains a component. If I wish to eliminate this from the entities. Therefore, I can now say the same thing in this line in this node. The subsequent option is straightforward. Therefore, the chamfer is one distance, which is three times one. Therefore, the distance listed on both faces is identical. Afterward. Currently, the three corners are chamfered at the same distance. If I wish to add one side and both sides in a distinct dimension, I will select two distances. Therefore, the initial distance, which is this one, is ten. Subsequently, the second is five meters. Additionally, azimuth and distance. "If I wish to specify." Therefore, the initial distance is ten meters, and the angle is 45 degrees. This implies that the distances are of equal length, and the subsequent distance will be sixty meters. OK, you can now observe the transformation. The final factor is breadth. The width is the distance between this point and this point. Therefore, the breadth is reduced. The number is ten. All right. Additionally, there is an operation known as offset. It is employed to establish an entity at a specific distance, for example. Subsequently, I shall enter. Additionally, this. All right. We have. Presently, proceed. The span will be relocated if I relocate it. Additionally, the line is acceptable. the line is joined to create a multiple slot acceptable. The circle is connected to the lines. The. All right. I will now just duplicate. Let me now select this image and observe. All right. Given that the distance is the most remote. You can now observe that we have generated an unnecessary condition.

INTERFACE PART 2 SKETCH, MODIFICATION TOOLS AND TRANSFORM

We now have the option to trim the drawing to eliminate the superfluous portion. Therefore, in order to comprehend this, I must first draw a circle within the line, ensuring that it is contained within a circle, and then add a line. All right. Now, if I wish to eliminate the upper and lower portions of the line, I can use the crop tool. Therefore, it is necessary to reduce two entities. Therefore, I will choose the core of this. Choose the entity that requires trimming or extending. Therefore, I intend to reduce the size of this entity. All right. And with which we aspire to alter. Therefore, this one. Therefore, only the higher portion is depicted in this manner. However, if I specify that I desire the central portion. Therefore, I will conclude that this is the case and proceed to damage it. Therefore, I am now interested in dismantling this entity.

Additionally, there are two layers, including this one. Additionally, it is divided into two components. Additionally, this command should be activated once more. Therefore, I am now able to choose this option. Additionally, this one. Presently, there are three distinct divisions. All right. I am now able to eliminate any of them. Therefore, suppose that I eliminate this. Suppose I have a line. Okay, I have a line. Additionally, I wish to extend it to this. Therefore, what are my options? I will navigate to the trim entities section and choose to institute trim or extend. So, in essence, this option comprises two commands. Therefore, I executed this task.

Additionally, the line is automatically extended to this location when I select this option. All right. Therefore, these were a few modification instruments that we can initially eliminate. To gain a better understanding of the mirror command, let's draw a few seconds. All right. Additionally, it is important to recall. And then, you know, I will draw the line like this. All right. What if I wish to place these two objects on the right side of the line? Therefore, I may employ the "I'm here" command. Therefore, with the assistance of the menu and selection. Selection and these two. I must now select the mirror plane, which is the default for this movement, and confirm that it is acceptable. Therefore, it is possible to modify certain parameters, including the x offset, which is 189, the y offset, which is 153, and the x angle, which is 270.

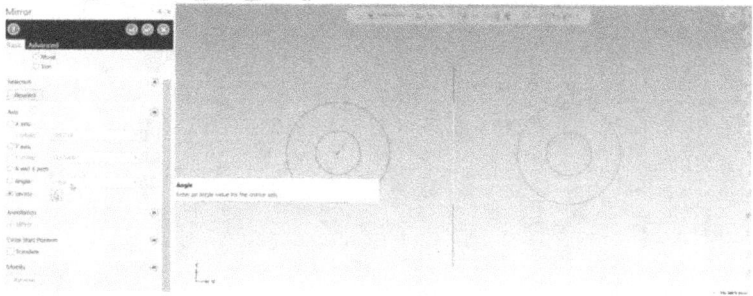

Therefore, you have the ability to modify it. Let us alter the angle while maintaining the values. Right now, I could be at a 90-degree angle. 1978 is acceptable. The same.

Let us assume that we require it immediately. It abruptly relocated to this side. You can now determine whether I. The year is 3060. 367 or 270. Once more, it occupies the same location. Similar to this, we have the ability to modify its y values. Therefore, the current value is 153. If I am to achieve a century, it is not far off. This exact same thing can be altered if I increase it to 170 and if I desire a modest variation. Let us increase the number to 200. All right. We will now proceed to the conclusion. All right. Therefore, this is the mirror command. We will now proceed to analyze this. We are unaware. Additionally, the radius. Therefore, I will select "Analyze entity" and observe that its radius is 48.75. All right. I am now interested in scaling entire objects to a value that will result in a radius of 50. Therefore, I can employ the "scale" command to achieve this. Therefore, that is the scan. I will now select and press Ctrl a.

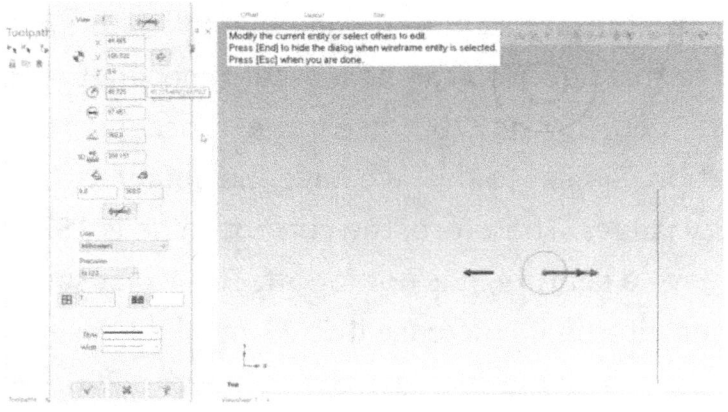

And the scale factor that we will employ is 50, as we aim to achieve a value of 15 when divided by the current value. That is 47 points. All right. So, we are currently presented with two alternatives: copy and transfer. Therefore, if I choose to choose a copy, the original object will remain in its original location. Additionally, the new object will be generated. Therefore, I will maintain my composure. I now have the option to "translate" if I wish to relocate to a different location. Essentially, the mobility of any access x I, y is restricted by translate. Consequently, if I am interested in only these two organisms, they are in this state. All right. In other words, it remains unchanged. Therefore, if I desire, I may create this one. I am aware that clicking on an item will cause it to be copied. Therefore, if I wish to relocate Scripture, I will duplicate it. Any randomness refers to the simultaneous operation of all three axes. Thus, it is completely meaningless. I can now establish its origin point and relocate it to this location. Okay, so this is similar.

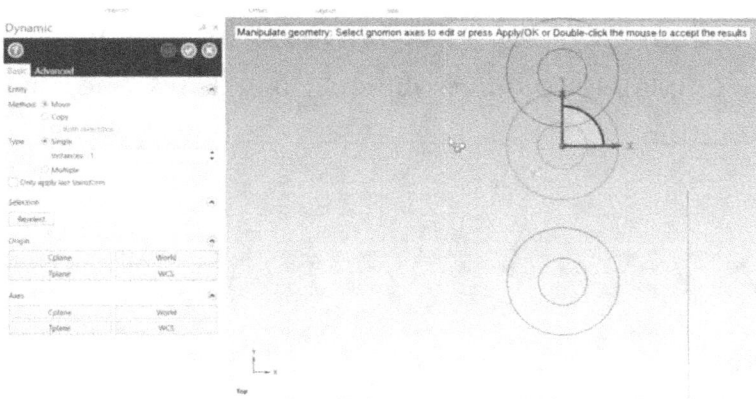

Place it in this mode and hold it. All right. We now have the option to retain the original sign and replace it with the new one. Navigate to the new state and execute the rotate command. I will generate a rectangle and grid. All right. I established a union in this location. Now, if I wish to rotate this. Therefore, I will descend into the afterlife on a specific date. All right. Additionally, the selection. I will now establish the rotation point. The location of rotation's center. Therefore, it serves as its focal point at present. Let us assume that I wish to rotate it from this point. Currently, we have the option to proceed in this manner. All right. Please maintain the temperature at 35 degrees. Additionally, seventy. False. Another example is this 145. All right. if I so desire. All right. It is acceptable.

False. It is an additional alternative for translating anyone. Therefore, if I select "translate," one will be generated. Therefore, the body that has been rotated will proceed to the right. Translation occurs when rotation occurs. The angle is acceptable. Another command is "roll," which is primarily employed to move objects, such as a stone, in a circular motion.

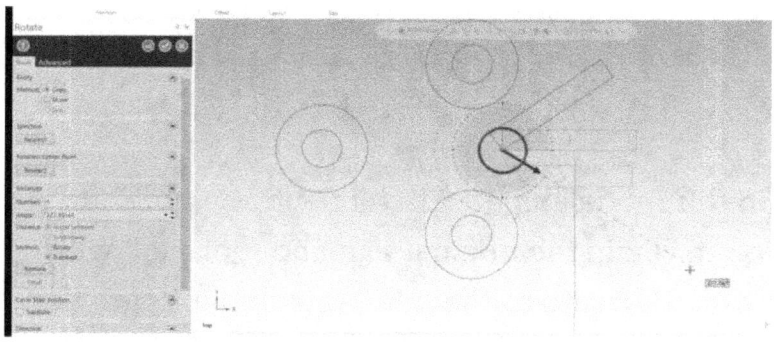

And I select this one, and you can observe that this endeavor has been implemented. Therefore, it is imperative that we modify our approach. Set it at a 90-degree angle. All right. Please observe the 90-degree rotation. Therefore, if we rotate this object by 90 degrees, we will obtain an object similar to this. All right. Let us assume that we desire to arrange a number of circles in the shape of a rectangle. Therefore, one potential solution is to create individual circles. Therefore, it will require a significant amount of time. Therefore, it is feasible to implement a rectangular array for this purpose. Initially, we will draw a circle, and then we will create a rectangle. Then the event occurred. All right. We desire these objects to be oriented in this particular direction. Therefore, I will navigate to the transform menu and select rectangular. Consequently, I will exercise control and selection. Therefore, you can now observe instances in which I direct one and I direct one are represented by the x and y axes. Therefore, we are interested in the quantity of instances. Let us assume that I desire five on one side. Then, two. It is acceptable. Additionally, the distance between them is proposed to be 5100. And in this, I have approximately fifteen thousand dollars. Currently, an area of the object is being formed. Therefore, this operation is extremely beneficial when it is necessary to depict something in a repetitive manner. Distribute is an additional function that we offer. In order to comprehend this, we will establish a boundary. All right. Exactly. All right. Therefore, I will now construct a circle at the endpoint, which is the angle one. Additionally, from this point forward, everything is satisfactory. Now, what if I wish to depict this in the same circles as the line? I require an indeterminate quantity. Therefore, I have the option of drawing a single circle and determining the distance between them. Consequently, which Ultimaker is it? Therefore, we have the option to distribute and transform. All right. Therefore, I desire for this circle to be distributed along this line. Additionally, I will select vectors on this page. So, the technique is to duplicate. For instance, suppose I desire five circles to be present on this line. Well, that's fine. Additionally, we have the option to specify the distance between them. Therefore, I have the option to specify the distance. Therefore, I can refer to distances. Fifteen is acceptable. And the utmost number is three. So the utmost number of results is ten, which is the only three. Therefore, 1212345 are generated. Consequently, we are unable to allocate additional capacity. Therefore. The limit has been attained.

SIMPLE CAD DRAWINGS PART 1

Therefore, we have addressed nearly all of the 2D commands and modification tools. Therefore, let us engage in some drawing exercises. Therefore, we will commence with the basic rank and progress to the minor complex. Subsequently, I shall execute a rectangle that measures 60 by 70 and extends to the far right. Afterward. I am in search of 7,260. All right. This and the act of contemplation and its integration into the focal point. I will now remove this line, as it is unnecessary. Arc endpoints will be implemented. Therefore, this is the initial argument. I will specify a radius of 35, as you are aware. The second item that is required is a circle with a diameter of 20. And I will click on it and attempt to locate the center fine line in layers, as shown. The center point has been obtained. The radius is 20. All right. This is a deceptive method. I will now demonstrate and position the cursor on this point at a later time. Subsequently, I will move the cursor to this point to ensure that it is the midpoint. Exactly.

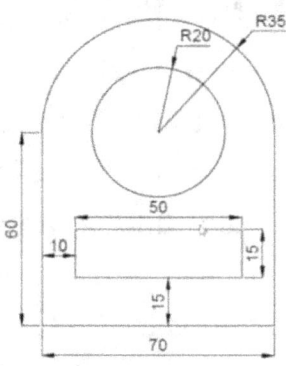

Subsequently, the rectangle of 15 units in length and 15 units in width is presented. Therefore, we can depict it by employing a center. Therefore, if the value is 15, the center will be 7.5 15 plus 720 2.5. Therefore, I establish a line in the center. Twenty-five. The subsequent item is 20.4, which you can shift by moving 1025. In this manner. I would like to select this point. Additionally, I am at a 90-degree angle. Additionally, 22.5 to. All right. Therefore, I will now form a rectangle. From a total of 50 crosses, 1515 crosses were constructed. You are in good health. Anchor yourself to the center. Thirteen and fifteen. Our drawing has been finalized, as you can see. Therefore, we proceed to the subsequent triangle, which is somewhat more intricate than the previous one. To begin, we will draw a rectangle in this location. Therefore, it is 70 in addition to 100 in addition to 70. Therefore, the width and height will be 214, 15 1650, and 116, respectively.

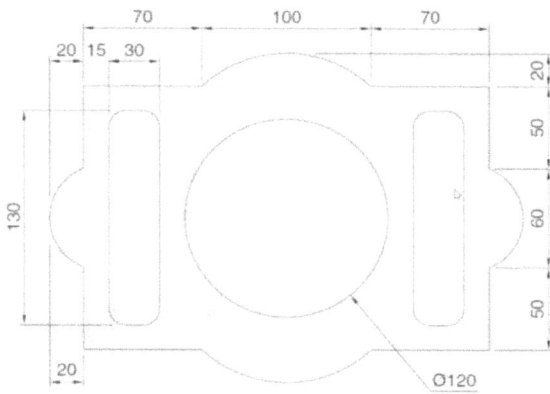

Therefore, 240 and 160. Therefore, we will implement it in the interim. 43,160. It possesses one. It is then time for the subsequent operation. The sole action I took was to ensure that the final line, which corresponds to 120, was the center. Therefore. It is the central focus. Twenty meters in length. All right. Additionally, proceed to relocate. Therefore, I am deleting this file. It is not addressed. Therefore, I will take it into account. Additionally, I will evaluate it as five. Therefore, the central point is now required. Therefore, I proceeded from the center, resulting in a value of 15.5 out of 15. Therefore, it will be thirty. So, I will proceed to establish a line from the center. To the age of thirty. Thirty and more. Commencing in the year 1009. I will now generate four rectangular shapes. Origin is the option you choose.

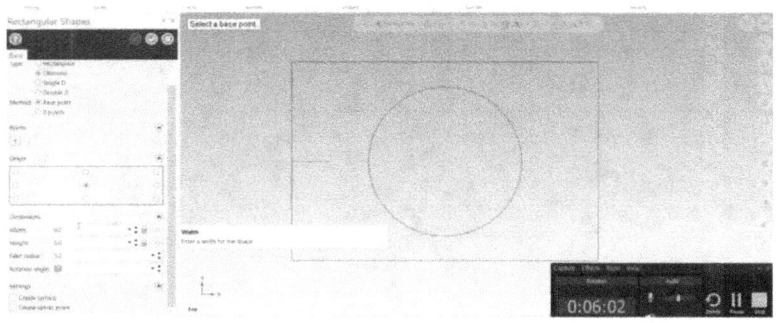

Then, on the round and designated uses, let's say five, and what will be what will be 30? Additionally, I completed 130 tasks. To the age of thirty. I succeeded. You are aware that you simply click on this point and you are fine. Subsequently, the for is decreased to 30. I am in the process of searching. All right. Additionally, this one. Afterward. I will now proceed to develop this section. And then I will center on both sides. Therefore, we will provide you with the most recent information. So the distance between these two points is 16 and it is presently at this outer distance. This outer point will be selected by offsetting this line by 20. Therefore, for this offset, I and 20 are acceptable. So, I divide this line in half. Afterward, you select this option. I am it. This and this. Therefore, we have successfully identified the central point. In reality, the duration is unnecessary. So, you simply require. All right. We will now proceed to draw.

The midpoint and. I will set it to 60. Face 60 and this series. Okay, sixty. I will now employ arc endpoints. On the contrary, this one and I. All right. So the radius is 32.5 in this case. All right. Presently, there is no requirement for five. Therefore, in order to establish a connection, I require the following: this name, this, and this. Additionally, quantities. It must be permissible. Additionally. I will proceed with this line, and when I am admitted, I will require that line and center. Therefore, 88 and one are centered. All right. I will now proceed to the transformation. Then, minute one. Therefore, this is the case. From the community. Additionally, there are items. Afterward, we acquired this cranium. Therefore. You are aware that we must depict this at a height of 20 and a scale of 100. You also perform the same action in this location. We will now proceed to run this center line from this center to this center. Subsequently, we will turn left and select on "okay." And we have a central point. Therefore, I will establish a line and a point in the middle of the column. Afterward. Is this line. All right. Once more, proceed to the endpoint. This line will be offset and dimmed. I will now proceed to the utmost points. Additionally. Will not eliminate superfluous lines. Including this one. Then, the second line and orientation. Additionally. All right. A central path will be established. Subsequently, I shall execute the transformation.

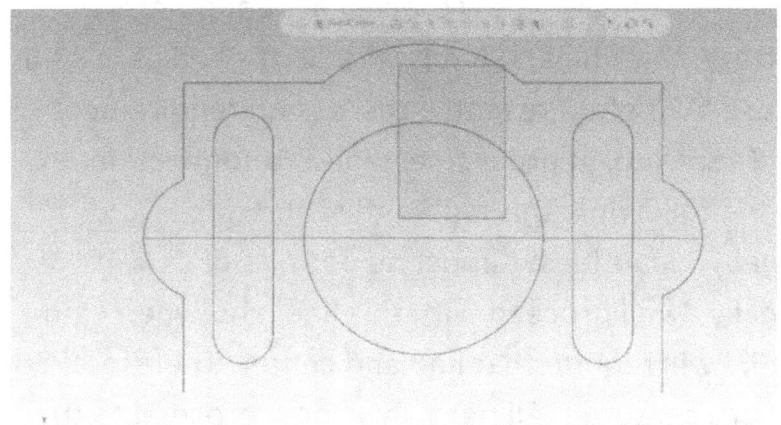

Additionally, there are instances of this. I am going to choose this. Additionally, the selection. This may be acceptable. Subsequently, we shall align this matter. We have our design. Perfect. Therefore, I will initially generate a line of 129 and its endpoints from this drawing. We will trace this circle within this pentagon. Therefore, we should proceed to assign it a name. This is the case. Additionally, I believe that we are required to include a frame in a 120-foot line. We should reposition it. All right. Therefore, I shall allocate. All right. Currently, we are attempting to determine whether the value is 129 or greater. Therefore, the entity is progressing at a rate of 29. We will proceed if it is not. Therefore, in any case. Therefore, the radius of 1300 is greater at the line's terminus. The current time is thirty. Subsequently, retrieve it. This number is nine. Subsequently, twenty is reached. I will now draw a pentagon with no internal hexagon of 20. Therefore, I will proceed to the conclusion and volume side six radius here. Then, in. The.

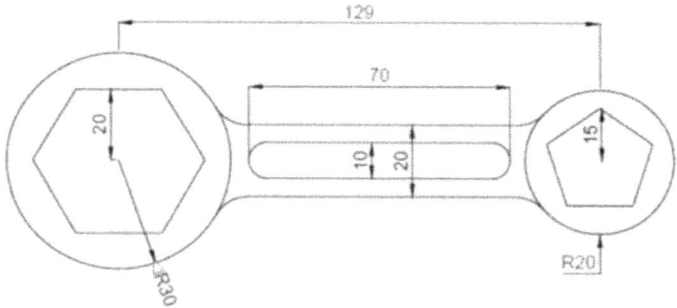

Presently, this is the case. Afterward. All right. All right. So now that we have five sides, it is a. Radius d. Additionally, and. Select "OK." Currently, we require a position pass with a breadth of ten. Additionally, there is the matter of height. Therefore, I will proceed to rectangular shapes in brown and bind them to a width of ten. Height: 17 inches. Width: 17 inches. Additionally, you specified a radius and height of 70. The location and landscape of this location. Moderate environments and. Therefore, it is not equivalent. All right. It is disseminated. It is. Therefore, we should exercise control over this matter. And if we have a small function, then. Therefore, it is necessary to reduce this line initially. Trim this green using this and never trim in this line using. From time to time, engage in simulations. Additionally, you have the ability to transmit. The same height. You are aware that I will offset these lines by ten. Therefore, proceed to offset. Distance. Next, construct it. One, two. Therefore, it is twenty. Therefore, it is ten. From the center. All right. Additionally, choose. This is the one. Also, I believe we are accurate about ten. Once more, proceed to the offset. All right. A distance of ten is acceptable for both sides. Afterward, additionally. Execute this action, and I am currently engaged in it. Afterward. Clicking on this will enable me to connect these lines. And this radius is five. Therefore, we will maintain its inclusion. We are opposed to trimming this. Additionally, this is not adequately defined or integrated. This can be altered in a similar manner. All right. Alternatively, this line is entirely relaxed.

SIMPLE CAD DRAWINGS PATT 2

Therefore, we will now proceed to a more advanced or intricate training. Therefore, in this training, I will initially generate a line of 100, generate a circle with a diameter of 30, a radius of 30, and a similar radius of 15. Additionally, it is forty. All right. Therefore, we should commence with the lineup. Consequently, the extent of the data line is two. Additionally. All right. Additionally. Add this to the equation, and I am required to generate a circle with a radius of 13. Additionally. Therefore, the radius of. And is thirty, and that is. Nine meters. After this, we will have a circle with a radius of 30 and a diameter of 15 radii. Therefore, and. Then, it is 14 degrees to the right and then over in the 15-degree radius. All right.

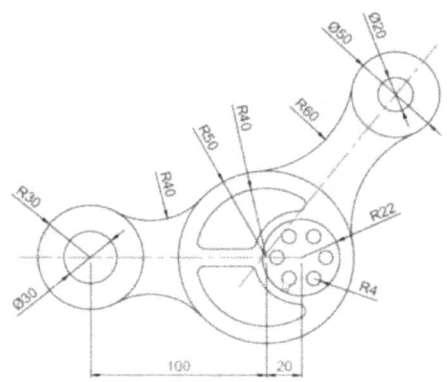

We will now generate an additional 20-line segment and a 20-degree radius circle to ensure that they are in alignment. Click. Also, release 22. All right. Currently, we possess an arc with a radius of 14. Therefore, I will employ the circle edge point for this purpose. It is also possible to employ a tangent arc. Therefore, it is fourteen. 17. Initially, I intend to retain this segment; however, I am not interested in retaining the other. Ease. I appreciate it. Accidentally, I. I am aware that this is a 14-point arc. Therefore. All right. Its values. Therefore, I will establish a value and arctangent for this. I will choose our two entities because they are tangent. So, the radius is 43 minutes, correct? We should review this. And I am now inclined to maintain this one. I now had an internal operation. I am now inclined to distribute this item. All right. Therefore, it is evident that the train does not specify the angle. Therefore, we shall execute it. Ah, well. Additionally, it is exceedingly straightforward to execute one instance. Therefore, there is no issue. I will now generate this small circle with a radius of the blue circle, as we do not have an inverse for this. Therefore, spin. I am capable of extracting the fasteners.

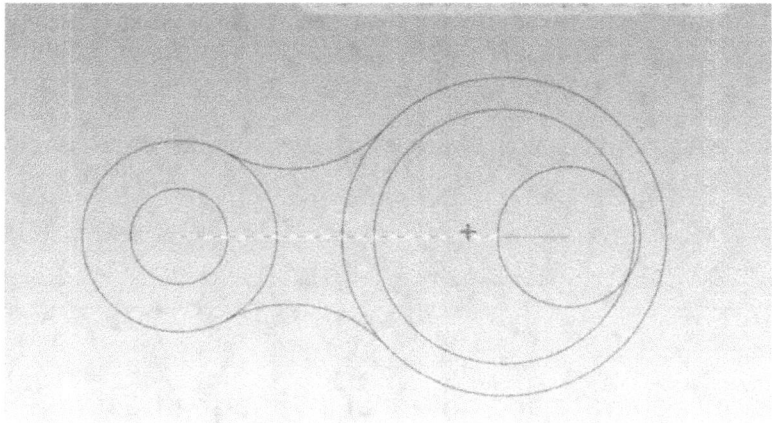

And one and 26. We possess six. Select this location. Additionally, we possess six seconds at this juncture. Additionally, the diameter of. Leg 22. I must, and we must have the arcs of. We do not include arcs, or it is only 28. Prolonged. Therefore, the year 2222 will be nearly an. I propose that we utilize the number 30. All right. Thirty is. To offset this circle, I will. Three years. I use the term "offset." The designated jurisdiction is subject to 35 restrictions. All right. Additionally, displacement. Chosen. Absolute values and saved. Will be the sole. Additionally, I will establish a line that is inclined at a 90-degree angle. All right. Extend. All right. I will simply provide a detailed explanation. Thus, from. The time is satisfactory. Therefore, it is incorporated. Unfortunately, the lens will result in a lack of length. Let us speculate that the number is greater than or equal to 16. Twenty-five. We previously had an office that was 15 by 110 feet. Additionally, this item has a diameter of 1525. Additional to ten, there are 135 and ten. All right. Additionally, 35 is the length of the lenses. Additionally, I am capable of decreasing 3060. For myself. All right, all right. We are now at the conclusion. Matrix. Therefore, I will initially generate a circle with a diameter of 2050. 9 to 20. Oh, it is due to the fact that it. Additionally, there are 910. Fifty. Currently, the tangent of the radius is half of 60. All right. Afterward. Entities. This is the 61st radius. We permit this individual to distribute it. This is the case. The object is not particularly near. I will make this minor adjustment. We are unaware of its name; therefore, I will make an educated estimate.

First and foremost, I will assert my ownership of this region of the circle. Therefore, this will endeavor to. I will assert my ownership of this. Afterward, this occurred once more, followed by this. Sure, let's do it. Relocate. Additionally, in this manner. All right.

I have an offset in this line for this section. Distance and a ten are present on both sides. Yes. This should be repeated for a total of ten. All right. We should fabricate. It. Additional. Afterward. I possess only one. I will shatter these boundaries. It and. Furthermore, I am shown to be extremely unfit and to have violated this principle. Additionally, at this juncture. I can now relocate this line. And a cylinder to this. Additionally, this. All right. Additionally, we do not require this one. This is where we shatter it. This is the purpose, correct? Therefore, I execute it by applying this force. In the same vein, we refrain from departing this side. Therefore, I will proceed from the beginning. All right. Break this one at this point. And then this one. This is the one. Therefore, this is the next step. Additionally. This is the case. Two. At a later time. The verticals. Line. All right. I believe that this is the case at this juncture, if you execute this line. All right. I will now generate a silhouette using this offset and this one. Additionally, this one and, and it is either one or trim. And indeed. In this manner. Similar to this situation. Additionally, this. This will. Arrive at this dismal hour. Therefore, I will collapse. This individual records this information, which is followed by this. This one is comparable to the other and will be similarly necessary in the current state of the game. Delete it. All right. I will simply reduce these. I intend to shorten this line by employing this. Additionally, it should be trimmed. All right. The window was broken by this line, which I have intended. In fact, it is possible to employ a pruning interval. We will maintain this exact same function. However, it is necessary to delete it separately from this or. No, no. And from the I mean reduce, I would like to use this line. All right. Therefore, it is essentially eradicating this one. Therefore, you simply destroy it. You are aware that we have the ability to remove this post. Therefore, in order to accomplish this, we should eliminate the center line. Then, I become more stringent. This is an action that we took. Additionally, we possess this inner community. Additionally, we are lacking in this ultra-push. I can either remove it or trim it. Therefore, we should proceed to this point, and this will conclude the matter. Additionally, this was not the case. It is permissible to eliminate this. Perfect.

3D DRAWING PART 1

Therefore, we should proceed with the section, which is nearly finished. Consequently, we are now proceeding to the results. Therefore, we will commence with this. Therefore, in order to illustrate this, we require this illustration. We will commence with a rectangle with a width of 6160 and a height of nine plus 12, plus 918 plus 1230. Initially, we will create a rectangle that is 60 inches by 30 inches. Additionally, we will generate circles. and this of the subsequent item, chamfers. Therefore, we will navigate to the wireframe and create a rectangle with a relative length of 60. Therefore, the breadth is 68 on the right. All right. Then, we will tend to center our attention and select one that is appropriate for the situation. Afterward. All right. Therefore, we currently possess this region.

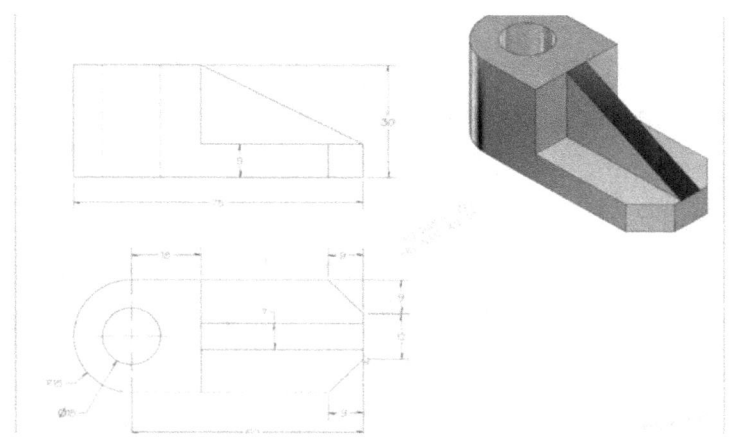

This chamfer is characterized by an equal distance of nine on both sides. Therefore, I will proceed to Chamfer and walk a distance. All right. Therefore, it is acceptable to trim it in the manner of this and this, as they are essentially the same bottom tip. Now, the second operation will be to offset this line by 18 distances. All right. Offset. Like this line and this height, which are a result of the distance of 18. Currently, we possess a circle with a diameter of 15 and its center. Therefore, the circle at the center is acceptable. Advance from the center position. Additionally, the meter is 15. Currently, we possess a radius arc. All right. Therefore, I will proceed to the arc endpoints, where I will circle the edge point. Therefore, the initial point will be this single segment, with a radius of 1215. Additionally, twelve. All right. This line is no longer required. Therefore, we should eliminate this line. Additionally, we require a line at the seventh position. Therefore, in order to draw this line, I will establish a line at the center. Next, I will rotate it by 3.5 and offset it. In order to determine the termination locations, we will draw a line from the center of this line to the center of this curve line. I will now offset it. I will adjust it, as we do not require a center line. Additionally, there was no change in the distance of 3.5 on either side. Therefore, choose this line. All right. We are currently attempting to determine whether this is acceptable. We will now proceed to the new section. Therefore, let us ensure that there are some reliable operations.

We will provide a practical explanation for each of them, such as a specific object. So, the initial operation, which is evident, is the extrusion, which is nine. All right. I am presently in a position to observe, so I will proceed to do so, and I am a combination. It is isometric. We will now extrude this portion to the nine and then. Therefore, extrude essentially generates a solid from the specified design to the specified height. Therefore, that is satisfactory. and extruded, go to the designers. Extrude and select the angle with this chain, and then attach it to this ten.

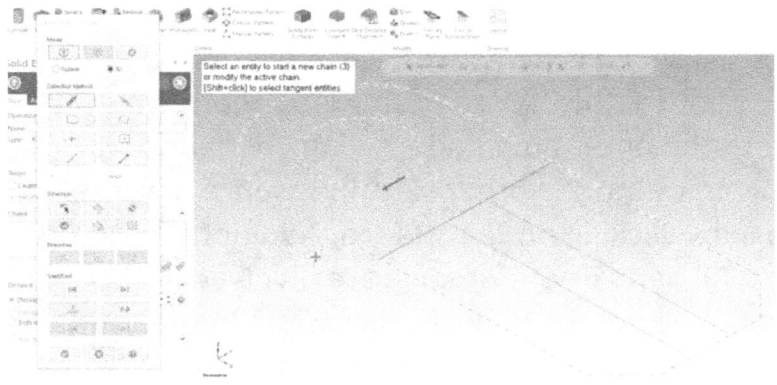

Additionally, we are excluding the ocean, which is acceptable. Therefore, we have chosen these two chains. Therefore, the region between these two chains will be extruded. The entire situation remains satisfactory. Therefore, it is imperative that we verify its accuracy. Therefore, it is either accurate or permissible. No, I have not specified the distance, which is nine. If it is feasible to reverse its trajectory. similar to this, but acceptable. Five is the answer. Additionally, I will respond at a later time. performing an additional operation. All right. Currently, I would like to project this small illustration onto this side. Therefore, what are my options? One activity that I am capable of performing is project management, while another is exceedingly straightforward. I am capable of composing this; however, we are unable to do so. We should incorporate some of the back into two separate sections. This line at this point and this line at this point. okay. Currently, this is the line that must be followed if we wish to decline a portion of the portion. Additionally, I believe that this is this one and this one. Exactly. and it is not, which is why it is defective. Therefore, let us conclude for this, this, this, and this. And then I must proceed to the one-frame undertaking. All right. I will now project it onto the other side. In essence, I will select this visage by clicking on the surface. Additionally, the selection. All right. Therefore, we are involved in this endeavor. I will opt for a specific occasion. In the event that I wish to affiliate with these lines, I will select "join." Additionally, this maneuver is acceptable. This is a composite. And one of the driver's limitations. It is excellent, yes. Currently, we have only projected the necessary portion of the sketch onto this. I will now proceed as usual. Therefore, select and extrude this. And in this scenario, you are aware of the amount of extrusion we have: 30 inches, which is equivalent to 21 inches. This time, the 21 vertices are acceptable. All right. We now intend to diagram this section. I will once more employ the command. Thus, it is employed on an almost continuous basis. Therefore, you comprehend this directive. The. Therefore, I will choose this line, this line, the wireframe, and the endeavor. Alt plus is capable of rotating this base, and the selection of the normal line yields satisfactory results. I will proceed to the solids extrude process. So, we have these two lines. Therefore, I will proceed to the wireframe. This is the point at which the line is formed. Similarly, it should be aligned. Okay, so I have it out here on the right. Currently, it is in the process of establishing a chain. Now, if I select "extrude" and choose this, it will. All right. I have drew it twice. Therefore, it is acceptable. Announcing. And extrusion has already been chosen. However, the second operation that you were required to perform was a chamfer on the distance. Therefore, we have a profusion of three alternatives. Therefore, let us determine which option is selected as the starting point. Therefore, it is essentially -18°F rather than 60°F. Subtracting 18 from 60 yields 42. Therefore, the length is 42 and the height is 21. Therefore, I must proceed to the distance himself and select this option. Okay, I will create a visage. Additionally, there is a choice. Therefore, distance one will be the complete length, which is 60 -18. That is forty-two. Then, twenty will be added. All right. Therefore, our final section is as follows. Additionally.

3D DRAWING PART 2

So, we will now rehearse this section. Therefore, in order to illustrate this section, we will first create a rectangle that represents 110 plus 18. Therefore, it is imperative that we commence at the appropriate moment. An vacant rectangle from 110. All right. All right. The format is as follows. All right. The second requirement is that we can chamfer it in design mode and within the program. Therefore, I believe we will proceed with concrete modeling, as we have already completed it in design. All right. I will now proceed to develop this section. Therefore, it is essentially a 40-cross with a 25 at the center. Therefore. Twenty-five photographs. All right. I must enter this and will eventually reach rectangular shapes. I will now select the origin in the following manner in order to draw a rectangle. All right. White. Is this the case? Additionally, we have a zero-sided return center for the rectangle. Exactly. Therefore, the width is 40. 14. And the height is three. Therefore. Right now. All right. The next step will be to solidify the material and determine the amount of extrusion. Here is twenty-five. All right. However, I neglected to incorporate the circle.

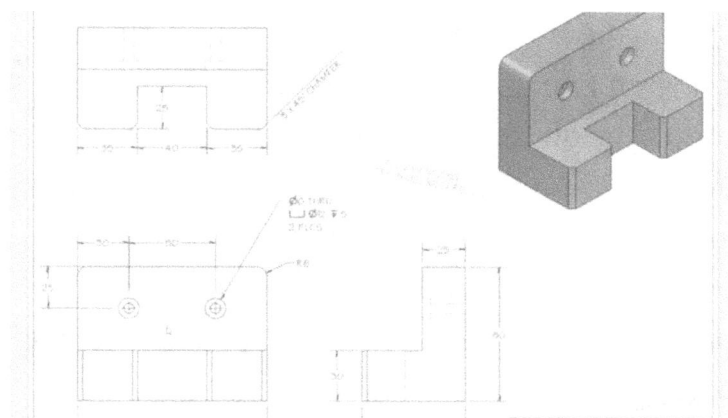

So, we can incorporate these circles, which have a diameter of six, and the ones that are sixteen, five, and six. Then, that is of a diameter of 12 and a distance of 25. This will be accomplished by first offsetting this line by 25. Therefore, we will proceed to the wireframe offset. Create a copy of this line by setting the selected side at a distance of 25. 911 is sufficient. Therefore, the amount of 30 on the designated side this time. Eight. The offset should not be limited to the side of the center point. This collection comprises 15 items, each with a diameter of six and ten. All right. The same will be true for six and eleven. Then, at the same time. The diameter of. You will. All right. I can replicate this onto this point plane and achieve a transformation. Additionally, I will employ the translate operation. okay. We are aware that the aircraft can be translated and rotated. I am simply drawn to this one, this one, and the selection, and I will proceed by trial and error. Yes. Copying is the method. Therefore, I made an error. Clicking on France will result in a subsequent selection and click on the right arrow. Completed. Ordinary. And culminate in solid extrusion. This is the tenth. Therefore, there are two. Device for pointing. Therefore, we have a six-inch diameter that is one through one. Therefore, I will opt for that. This area will now be extruded, but I am unable to observe it.

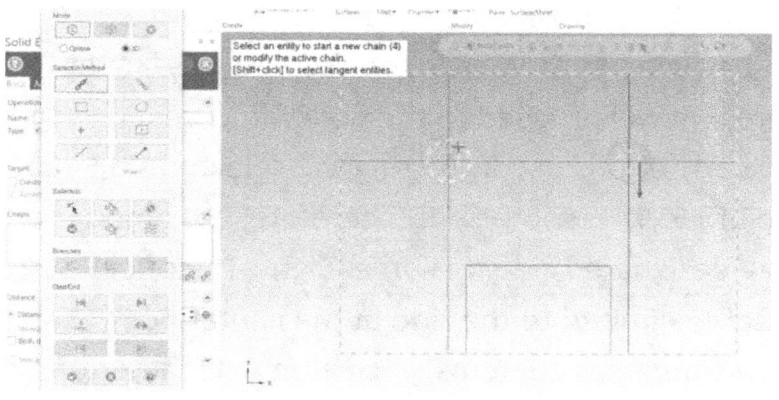

I proceed to observe the metric, which is the distance, which is 2525. All right. Currently, the undertaking has been discussed in the previous project. Therefore, I will simply reverse its direction in order to facilitate the subsequent procedure. Additionally, it generates planes. This is unnecessary. Or, at the very least. We are now going to extrude. I will eventually reach solids and is. We possess a line crater with a profundity of five. Therefore, I have extruded and selected the peripheral items. I will now select Operation Cut. Additionally, I am exceedingly rigorous in this location. All right. Therefore, we have now established openings. Sufficient. Therefore, in order to administer this individual, I will establish a rectangle that measures 110 by 25. Therefore, proceed to acquire a rectangle. And for this, they arrive. Change the number to 1020. All right. Presently, employment is satisfactory. Select it and extrude it. Select the chain and select. What is the reason for the failure to select and execute again? Is it now acceptable? It is permissible to generate 41. All right. All right.

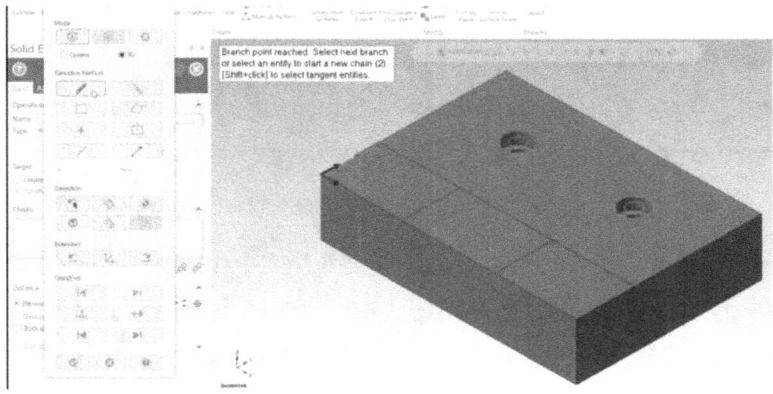

Therefore, we are required to extrude this rectangle to the specified distance, which is absent from the given information. Will manifest at the conclusion. However, we do not possess its dimensions or the location where X discovered them. All right. Therefore, its dimensions have not been specified. If we have a distance of 30 from this node, we can simply append it. Therefore, it will be nearly 367867. Subsequently, it was extruded to seven. Also, select this item within this chain and extrude it. The operation is the creation of a body for the card. Additionally, adhere to the direction until it is satisfactory. Please ensure that these seven items are completed. Simple enough. These two are now required. Therefore, these are essentially rectangles that are 35 inches by 25 inches. Next, we proceed to the wire frame. Thirty-five. So, let us deliver. It and its width are both 35. Additionally, the street. Heading and derivation are provided below. Therefore, this. In this instance, the width will be reduced by ten. Additionally. We have chosen the opposite of "okay." So, this is the extrude process. Select this and this and extrude the two required components. The total is 6025, which is equivalent to 722, and 69 is 32. Therefore, it should be expanded. You are 28 years old. 28. There will be eight. Well, that's fine. In this case, it is possible to rotate it. We must now perform some chamfers. Therefore, we had three pillars with a radius of six and a center of five. So, let us arrive at the mastery level and execute it. Therefore, it is imperative that we occupy it. This one and this one. What it is: six. Additionally, it is six. Yes. Presently, we possess ten frames comprising no. In my opinion, these three cross designs are acceptable. Therefore, I will navigate to the contemporary and distance options and make my selection. All right. Next. The subsequent method of selection is edge. Therefore, choose this one and then this one. and this. We. Distancing ourselves. A 2D angle of 45 was used. All right. Exactly. Drawing is permissible. Indeed. Subsequently, I executed a sketch to examine the train mode vector. Therefore, in order to view this for me, you must navigate to the "hide" and "unhide" sections. I would like to retain this. This is. And then, hand selection. Therefore, presently, there is no illumination.

3D DRAWING PART 3

We will redraw this drawing using a straightforward command that includes cylinder blocks. In accordance with. Therefore, we have already generated this model by employing extrude. However, we will now employ this. Therefore, I will initially illustrate this 75-by-30-inch rectangle.

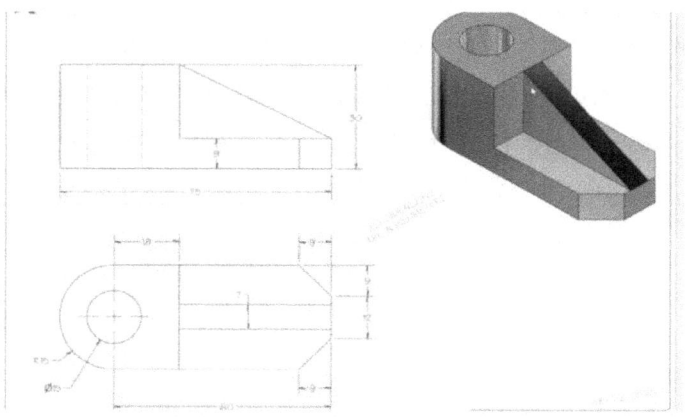

That is fantastic. Therefore, I intend to generate this upper portion by blocking any point and 7530, and then N. I require an additional 21-block. And 21 of this, which will be 32, is acceptable. Block and click on this, select it, and bring it in. All right. Alteration is also possible. Therefore, the length will be 32. The width will be 30. And the height will be 21. All right. I require this zero that we discovered. Therefore, I made an additional block in the center. All right. Therefore, obstruct. Additionally. The central location. However, this window. Therefore, the dimension will be seven. Cross 16 is comprised of 18, 42, seven, and 42. The following numbers are 20, 42, and 27. This is where its nucleus will be located. All right. The sum of seven and eight was 42. Twelve or more. All right. First, we will fill it in. So, I will initially submit it. Similarly, this is the case over here. Selection.

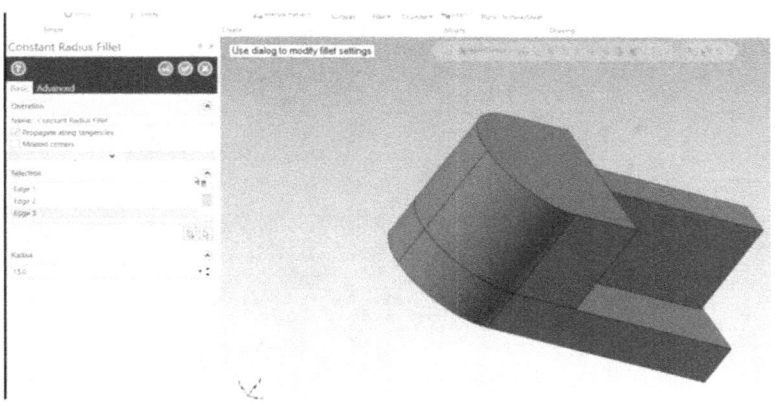

The number is fifteen. Then, I now comprehend the concept of transient. Complete this field. And this will occur in the. This is the one. Subsequently, in the distance, and. Exactly. All right. We must now establish a buffer zone. And then select "okay." I will then select another visage. Therefore, this is a single field. Two is acceptable for the initial distance. Additionally, this is the case. All right. We will prevail in this matter. We will now generate this body by employing the simple command block central. Therefore, I will initially generate a subset of 25. Additionally, it will reach 110 by the end of 16th century. It will have a width of 80 and a height of 25. So, one 8110 crosses 81. Additionally. Ten thousand eight hundred thirty-three distinct. All right. You wish to actually be able to. Additionally, we will relocate this 100. I will conclude this section if this is the case. Therefore, it is probable that it will be seven feet in height.

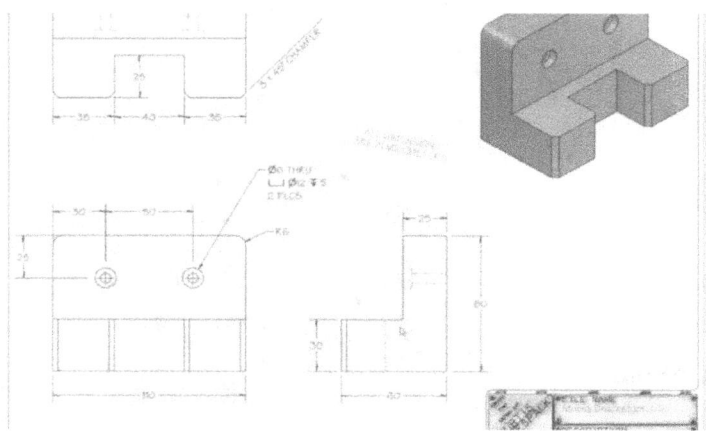

And in this manner, it is naught. It is comparable to 1025. So, it is a bit disorganized right now. I must first adjust the origin, and then we will arrive at it. Right here. Thus, it will be increased. Then it reached a height of 25. Let us commence. Then, it will be a choice between these two. Therefore, I will allocate 35 in addition to 25 for this purpose. The time required is twenty-five. Afterward, it will be. Thirty-five. Twenty-five. How it is described in this location. Additionally, sixteen. I. Let us assume that the number is thirty. All right. All right. Therefore, to obstruct upward in the mid-twenties. I will decide on this point and proceed. I must establish two layers, and temperatures are denoted by a loop letter. Initially, I will produce these openings. Therefore, these factors are all that is required to accomplish this. Well, that's fine. Then, I must utilize wireframes to construct it. However, I believe that we have already accomplished this. I am still capable of accomplishing it. Therefore, I must now establish a line. We would like to reduce the length of the line before generating it. All right. Presently, I possess this design. All right, I will execute the task. Therefore, this is where I must include it. This is the case. On this item, I am required to rotate. Therefore. Additionally, the selection. Afterward.

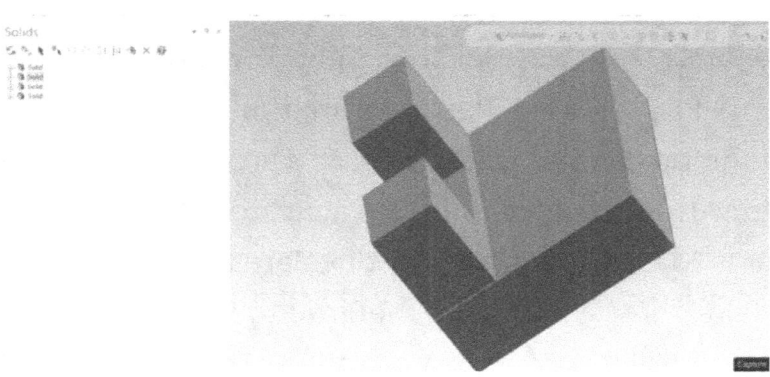

These lines will be offset, as you are aware. 25 and 3030. I will now proceed to neutralize this line. Additionally, twenty-five and twenty-five. Indeed. Therefore, thirty. For this and for this and for this, followed by this. x was denoted. All right. We now understand the objective. I apologize. The quantity and totality of the diameter 1216. All right. We are capable of reaching this location. Therefore, I will select the target board and the body. I will now select the complete variety that is most comfortable. Therefore, the diameter is 1200 gold. It consists of five Condon boards. Additionally, it is white. Therefore, the comfortable depth was five, and the number of layers was twelve. Additionally, there is a distance. I am required to include positions from this point forward. Decrease the rotation slightly. I am engulfed in flames. All right. It is the beginning, so it is imperative that we emphasize this point. Therefore, I am. Continuing this process, what is the purpose? Initially, we should establish our positions. All right. Location. Indeed. This item is located here. Positions are the sole subject. Enter. Additionally, this is equivalent to an 800% increase in position. This capability must be activated. The diameter is 612, and the left-most five are being counted. Additionally, throughout. I am required to implement this modification. Fill it with a six-inch radius. Therefore, I shall proceed to fulfill it. Select this as the edge and increase the selection radius to six. Six is acceptable. Yes. We are now at the position ring distance of one. It is imperative that we do so. Therefore, this one. All right. This is the one. Indeed. In addition to this. Additionally, the selection distance is three. Originating from. I am inquiring as to whether "ready" in this context refers to the correct sketch from the part that was not observed. All right. You are looking quite impressive in this location. Similar to this one. All right. We have now acquired knowledge and have drawn a single eight-by-two extrusion in a straight line. Methods that are straightforward. Okay, we will now proceed to the subsequent section.

LOFT ARC, SPLINE AND REVOLVE COMMAND

In this project, we will explore a novel instrument known as a ribbon, which is employed to generate circular bars. It is typically employed to create art that is symmetrical. In this instance, it is evident that this component is symmetrical. Therefore, in this instance, we are required to generate only this portion of the geometry design. Subsequently, the entire illustration is centered on, I am uncertain, that axis. Therefore, we will commence with a rectangle of three sides, with a height and width of 3.5 to 5. 3.5 inches in height and breadth. 3.53 inches in width. All right. Right now. Additionally, its breadth is not specified in the outcome. So, their shared victory and height are converted to height is. Three inches in width.

Exactly. So, the number is eleven in height. Additionally, breadth is. Therefore, this region and this are one and four, respectively. Consequently, two twos followed. eleven, two, four, eleven. Finally, we have seven crosses six.

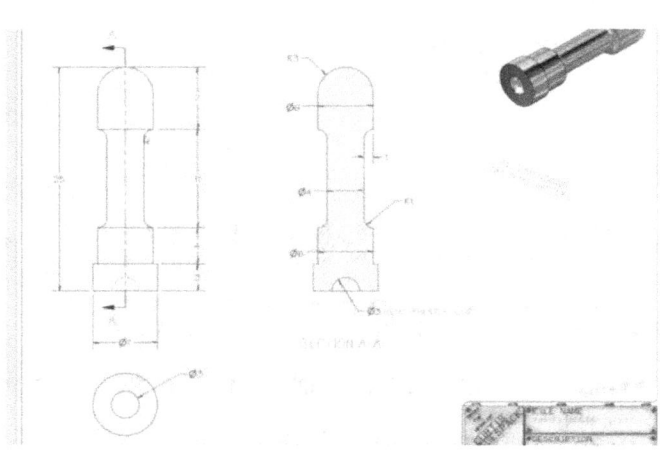

Therefore, the ratio of three to seven will be divided by half. We currently possess a significant number of threes, which is 15. Therefore, we should proceed. Well, it is. At the very least, two-dimensional. All right. A silhouette of one is depicted here. All right. Therefore, this is the situation. Finally, we have reached an agreement. Right now. These sentences are unnecessary. What actions did you take? You succeeded. Additionally, we possess a circle with a diameter of three. It is also possible to implement it through the use of arc or. All right. I will reduce it. Reduce this single aspect to this. I apologize. All right. All right. The quantities were satisfactorily reduced. Therefore, we should implement it using an arc. All right. Therefore, I will choose this point and perform the diameter. Therefore, it is acceptable to begin with the number 19. Switch it. The starting point will be zero. Afterward. From this point, the diameter will be three, and we will now rotate this section. OK, so please rotate this line and spectrum. Line. Trimming is underway for this individual. Therefore, let us create, disrupt, and execute. Yes. Afterward, this occurred. Yes. All right. Complete this task. You are responsible for this task. All right. I will now proceed to solids. I will return it to its rightful place. This one, this one, this one, this. All right. And here, we have the unneeded sentence. Five and ten. Eight. Rotate it. Remove this sentence. Let us achieve it. All right. Additionally. This is the case. Now, navigate to the wireframe by pressing M. and in addition to this, action. Do not attempt to fill in any voids. We will now transition to solids. And here, multitudes are present. Choose this option and click "OK." Additionally, on the horizontal axis.

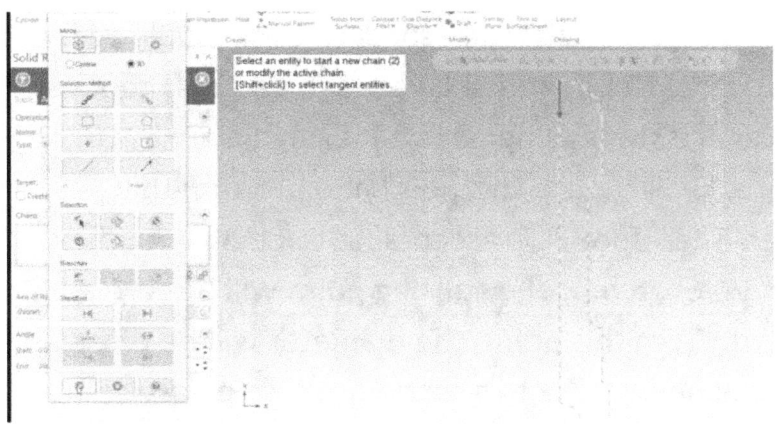

Choose a path. Therefore, this will serve as the axis. Therefore, if I am permitted to rotate. I will have you return home, and I am aware that they arrived with this item. Therefore, the revolve command can be employed to generate any form of symmetry body. We will now implement a novel instrument, which is also known as loft, following its designation. Therefore, we should commence. So, initially, I will navigate to the "View" menu and then select "Show Axis." All right. I then proceed to the wireframe and generate a radius and green silhouettes. And then I will counterbalance it by five units. False. I will right-click and select the isometric dimensions. All right. I must now construct an aircraft for this purpose. Therefore, I will select this item and relocate it to the center. This is the case. At this point, I will specify the value of z, which is eight. Additionally, it is fifty percent accurate. I will proceed by selecting "set is W C is okay." This has been functioning as a coordinate system. Therefore, we will proceed to construct it and subsequently assign it a name. I see. Additionally, the command. Right now, I will generate an additional circle within this plane. Additionally, it is fifty. It will be negated by me. Five is acceptable. I will now select the dog designated by W in the coordinate system and click on it. Select this option and then select "okay" once more. Consequently, a bell-shaped structure will be generated. Subsequently, I will implement dynamic rotation. So you can observe that, alright. I will return to my residence in order to locate it more effectively. I am. Click on "view" and "hide" when you are prepared. The axis is functioning properly. Currently, it is possible to determine the source of the image. Additionally, we have acquired a new command car for the purpose of sleeping. The primary distinction between slumber and revolve is that the former concentrates solely on recall and follows a circular trajectory, while the latter is capable of following any type of path. To begin, I will generate a path using the horizon. Therefore, I will commence with an. Within this. I will then right-click and switch to the front local coordinate system. Therefore, I will generate a rectangle. this. Then, I will locate the line. Afterward, it is acceptable. Therefore, it is necessary for me to project this line. Therefore, I will develop an endeavor. Within this and the selection, copies are distributed throughout the plane. This aircraft. Afterward, I possess this item. Therefore, I will now form a rectangle. Presently. Additionally, there are dimensions. Be consistent and select the estimated values in this section. Fine. At this juncture, it is imperative that we establish this point, this organism. Therefore, I will proceed to translate and transform. To this end. And then I proceed to this. I am aware that these are perpendicular. Therefore, I will proceed to solids and select "swipe" to switch to "sleep." Therefore, our objective is to eliminate this issue.

Then, choose the section along the jawline. Wonderful. Therefore, I would like to adhere to this. Ah, I see. We possess this. The segmentation. Therefore, this section is generated by utilizing slumber. Using the sweep command, I will now generate an elbow. Therefore, I will initially create a circle with a diameter of 15, say. 1540 is acceptable. Subsequently, determine the quantity of these two. Three units. There is none. False. This is the highest point. No, I will compose the blinks. All right. Therefore, it is necessary to establish a component. So, typically, I would use this element to establish a 90-degree angle. Therefore, I will generate a rectangle for this purpose. And from this point forward, how much do we desire an elbow? Consequently, we have the sum. The current time is 4:43:42. Additionally. Two 7470s. Now, we must eliminate these lines, as they are unnecessary. Do you not know, please allow this one and this one. The radius will be 30 meters. Additionally. I am planning to include two additional items. I will now proceed to solids. Therefore, let us investigate. Sure, we would like to eliminate these two. In this direction, glide along. All right. Therefore, this is our forearm. All right. We will now generate a straightforward scale by employing the sweep and cylinder commands. It is exceedingly rapid. Additionally, sustenance is derived from lambda. High radius. Additionally, I. Sixty. All right. I will now employ a 2D command, which is the I helix. Therefore, allow me to. Let us ensure that the remainder is level. The height will be acceptable, and the radius will be five. The height is 60 inches, and the sign we received is approximately this height. The pitch will be five, and there will be 12 revolutions. All right. I already have a plane in my possession, but I will demonstrate how to construct it. For the time being, move this to this point and provide me with this section. Then, proceed to the frame. Additionally, I will generate a circular at this juncture.

Therefore, I will ensure that my peers are satisfied. The diameter of this room is 4.5. All right. I will now proceed to a deep, uninterrupted slumber. The following is in this section. The body will be sliced by me. All right. For the time being, I will provide a brief example. I will maintain a distance from. Certainly. In the event that I. Okay. We should prepare it for. Yes. Because of the penalty. I will now include a cranium. However, I must first establish the values in this location. Therefore, I will incorporate a cylinder into this explosion. Additionally, the location. Afterward, I will be required to complete this task. Presently, we have established. Ignore it. Therefore, the entirety of this land is composed of strands. We can thread this section if desired. using I will select and it and. I will analyze the entity. Certainly, evaluate the entity. I will modify it here. Sixty. Presently. Consequently, we comply.

3D EXERCISE 1

This illustration will be produced by us. Therefore, we will commence with the line. If I. Therefore, we should proceed to the next step, which is to draw a line with the number 95. Additionally, the following serves as an indication. by means of which we will utilize this section. Therefore, we should navigate to the Master Camera and establish a line on a distinct line. Therefore, the midpoint is unadorned. Additionally, this sum of 195. All right. Additionally, we possess these two terminations. So, I will now generate these rectangles. Therefore, it is 100 times 55. Additionally, this item is an item number 1585. Therefore, I will adopt rectangular shapes. Furthermore, I will provide this component with an origin of 55 and a height of 100. Therefore, Iris and successively. This point is acceptable. Additionally, we. We will include an additional 85 units and have a total of 1015. However, this necessitates additional resources. All right. To this end, I will draw a line that represents the total of 175. Then, the line and it is. To render it as two and a point. Oh, so if I. I am capable of enhancing the patterns, you know. Click. At eighty-five and one. Currently, we are at the margin. Additionally, these two. 30 and 40. Thirty and more. Forty. Then, on this side, it is 5040. Fifty. Additionally, forty. All right. You are now required to execute the task. To accomplish this, I will create a rectangle with dimensions of 75 and 35. Therefore, if we divide this, the length falls within the range of 8255, -85, and 155. Therefore, the sum of 115 and 75 plus 35 is 710. Therefore, I will achieve a score of 77.5. And this 137 point is acceptable. Therefore, I will create a rectangle with a length of 70.

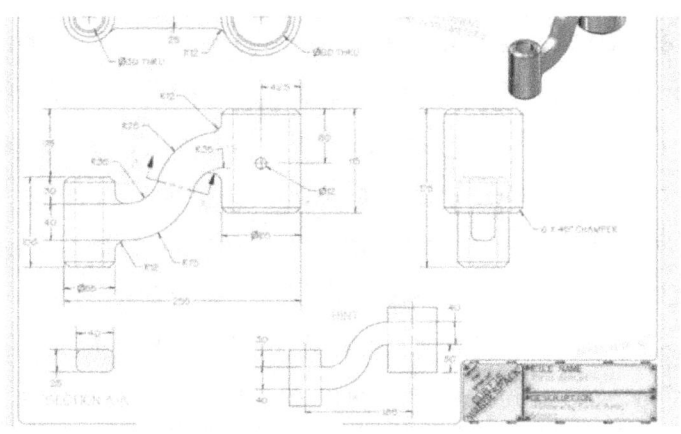

Nine is equal to seventy. Additionally, 7.5. And then, 77 and. In this instance and. It is once more my intention to proceed to the rectangle. The value will be approximately -77.5. and I will add -77, and then the rectangle is acceptable. The cost will be $3750. All right. This line is unnecessary; therefore, we should proceed to this location. This sentence is unnecessary. Eliminate it. Additionally, this one. Exactly. And transitions from one location to another. And we do not require this.

I will be entering diagonally. Next. So, initially, choose these two tones. 12. Certainly. Twenty-two. Additionally, the design is identical. Additionally, this and this are included. And we are 35. Additionally, twenty. We possess thirty. Also, this and this will be satisfactory. All right. We will mold. All right. And proceed to the subsequent line, and I will provide a list. All right. This will occur and. All right. I will now complete this and this. All right. Additionally, this one. All right. I have now acquired an offset. Right here. Chain with an offset. Therefore, you are presented here. We should disable it. Sources: here and here. I will now proceed to neutralize. Therefore, we and you, as well as the extent of our offset. All right. It is. Twenty-five. All right. This necessitates a power of 41. Therefore. Yes. It is forty years old. All right. It is forty years old. Yes. Additionally, it is evident that it is in the present tense in this 41-line passage. Afterward. I am aware that I will once again extend the lines and this to. Initial production. Which I am currently engaged in.

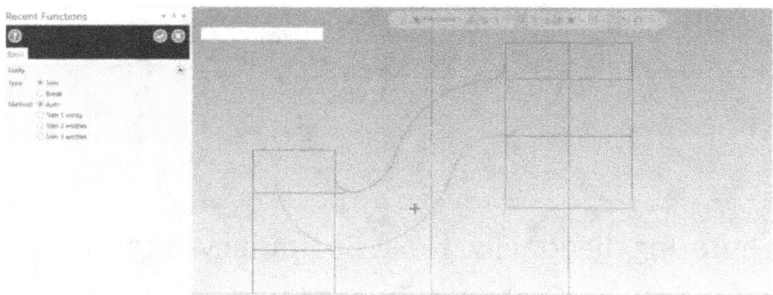

All right. Next. Complete the task. Reduced. Additionally. Therefore, it is imperative that we rectify this situation. All right. Therefore, we should. I will choose one of these two. Cylinder. Additionally, this one. This is the one. 12. Additionally, this one. And this sentence. We currently have trim stands in this line to this point and this line to this point. Additionally, as illustrated in this instance. Additionally, all that is currently in existence. Therefore, it is time to incorporate all superfluous lines. Therefore, this is unnecessary. We are interested in this item. Additionally, we require this item. Wireframe. Connect the endpoints. Additionally, this. Additionally. We should combine these. All right. You are assembling frames. Determine the distance from. All right. This is unnecessary; therefore, it should be eliminated.

Eliminate. In this manner, eliminate. Additionally, in this instance. Additionally, this one. All right. We may require this item. Therefore, it will be distributed immediately. Therefore, we should choose that. Additionally, the midpoint of this path. Subsequently, I will position my cursor at this location and time. And if I do, when will I be able to access this? All right. Therefore, we should initially modify the and it will be eliminated. It has already been placed at the top and front. Utilize a slender and this grid point, and I am fond of the 27.5-inch radius. Additionally, the height will be below. We should accept it. Additionally, it is isometric. All right. We will ensure that everything is satisfactory; however, we must alter its trajectory. Alter the trajectory. Completely opposite. Yes. Additionally. All right. I am required to regenerate it. Items of this nature. Subsequently, I shall generate this. Divide the length by two, commencing from this point and its radius, and the result will be greater than 15. Exactly.

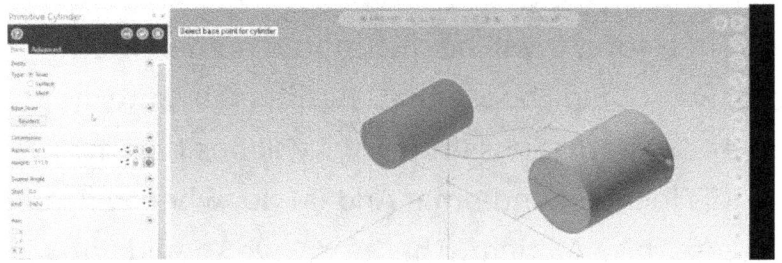

Now, remove these lines. And we have no further requirement. Viewpoints are employed in this context. Additionally, extrude. So, in this movement, it is OK. Additionally, it will be applicable in both directions. Additionally, verify that the thickness is satisfactory. The thickness is utilized by the technician. The thickness is 2525. All right. Additionally, I am included in the menu. I am correct. Proceed. numerical values. I understand. We require fillets at this time. Therefore, the initial step is the. Two chamfers at a 45-degree angle. So, incorporate the timber into the interior as well. So, the only thing that is outside is US 45. And now, I apologize; the sum of two and 45 is inside. Additionally, there are billions of them outside. All right. Certainly. Yes, it is six cross 45. I would say that's acceptable. Therefore, we require certain instruments. Therefore, I will do so for this purpose. One is the quantity. One is thirteen when it is sixteen.

Therefore, we should proceed to the holding area. Additionally. 15. That's fine. No, I am referring to the fact that we will create a wireframe and project in this, this, and this, as well as choosing this one. This visage. And we will make a decision. All right. The central lines are now in our possession. Therefore, we proceed to solids. Additionally, the location. All right. Take this position. The entirety is satisfactory. This concludes the matter. We will preserve this conclusion. Therefore, we should establish a distinction between each. Therefore, the initial step is to select and then assign a position. And with this one, and, we, and here we are. Thirty laminates were utilized. Additionally, there are two of them. Additionally, we should maintain it as a unit. For what reason? This is not taken into account. Therefore. All right. Regulates the. Control. Therefore, this extrusion. It is imperative that we perform this action once more. Cylinder. Therefore, the height is equal to the product of the radius and two values of 55. All right. I will now return to my residence. Additionally, that is acceptable. One meter of magnification is available, specifically 20. All right. Additionally, it is imperative that we alter our course. Two units and. Right here. Nothing. Additionally, in a diminutive position. Our objective is to achieve this. Afterward. And, and. Incorporate. Additionally, it has a diameter of 60. Additionally, we possess an alternative. Therefore, what is the extent of the information?

Therefore, it is six times 45. Additionally, this is six. Are you forty-five years old? Additionally, what is not readily apparent. Yes. Therefore, we must distribute it, and we will. Six is subtracted from sixty. Consequently, it will occur. Yes. The number is 66. Yes. It is satisfactory. Therefore, the action we took is 66. The inner opening measures 60. Also, this option was requested in relation to this diameter. Therefore, this upper. Therefore, we have provided it with plastic. Therefore, we should establish a single distance and angle. Additionally, choose. Additionally, this one. And the sixth option. I have it here in a relatively small way. Additionally. Yes. Furthermore, I will proceed to. One is acceptable. This concludes the matter. We obtained all of them at that time. Therefore, the same thing will be 36. Location. Two on the acceptable side. We should establish the definition of "distance" as 30 meters. Simply, I am a giver. The line meter is 36 meters. Refrain. Additionally. I will chamfer this one. Priority is given to this matter.

3D EXERCISE 2

Therefore, we will create this component at this time. Therefore, I will initially generate a rectangle with a width of 142, which is the sum of 50, 21, and 20. one. Therefore, the rectangle for the radius fillets is 142 cross 92, which is equivalent to 150 and 92. Consequently, I will return to the master's level. Additionally, I will proceed to wireframe rectangular shapes. All right. I will now reach the center. "And width was." The height was 92, and the value was 142. The radius of 92 is ten, which is acceptable. 142 individuals successfully completed the task. We will now generate an additional rectangle of 150 square meters. Additionally, anchor the anchor to the center. Additionally, there is fifty. Therefore, the objective is to generate openings. Therefore, there are gaps which are acceptable. 18 millimeters in diameter. Alright. It will have a diameter of. Then, input it into the system. Additionally. All right. Therefore, it is feasible to eliminate this item. Eliminate. Eliminate. Additionally, eliminate. And I will simply move to the right. I mean, that is preventing. Control P. Therefore, we have a region that is circular in shape and select it. False. Fourteen meters is the height. I will modify its operation. Consequently, this design occupies the foremost position. Consequently, it will be effortless to generate an additional motion on that subject. Additionally, we will generate two rectangles. One of the fifteen females is 192, while the other is 41 plus 41 plus 15. All right. Therefore, we should generate these and proceed to the wireframe rectangles. Place one rectangle in the center. Also, the width will be 15 and 20. We are ninety-two. 15. Additionally, include an additional item. It is informing me that the sum of 41 and 41 plus 15 is 150. Then, I perform fifty. All right. Therefore, I will disregard this and proceed to select the channel, which has a distance of 58. Additionally, operation and. This will be the 41st. For example, we will modify the chamfer to accommodate a distance of 22 - 58 -22. Or fifty-eight. -92 is acceptable. Additionally, this alters the distance. I am 41. We will now confirm the accuracy of our actions. Therefore, this distance must be 22. It is acceptable to be 22.

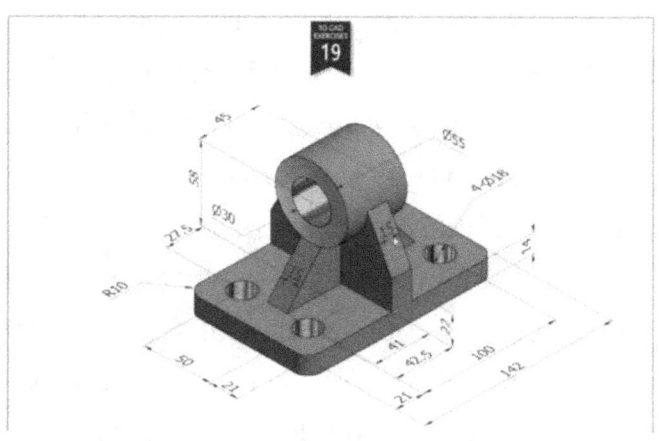

Additionally, this distance. 41. Yes. The number is 41. Exactly. In order to illustrate this one, we must establish a planet at an offset of 58. Therefore, I am in the process of establishing a. Hold it in. Yes. And relocate it to this one. Yes. This is comparable to the original. For a period of time, this initial number and this or this. Therefore, I will proceed to the height of the individual, and I am solely interested in this item. All right. Now, I will formulate a strategy. Additionally, Z will be 58 years old at this time. The plan is to write w.c. and create two circles, one of which is 30. Therefore, we must now create, and not 33 and 55. And then, thirty. Then, 55. All right. isometric. Its width is 45. Therefore, we should proceed to solids after extrusion. Initially, we will unhide and. Yes. We should fabricate. Therefore, we should extrude this and be in agreement. In both directions. Also, it is forty-five. All right. The reason it is 45 by two is that it is being translated on the incorrect side. All right. That is satisfactory. However, the necessary action is for me to enter. All right. The operation is canceled, and it. for this board direction and for this, and I do. Perform nine rotations. If we observe that the interior portion remains. Therefore, in order to eliminate this, we will implement a shearing operation. Therefore, which will it be? I will proceed to extrude and select this option. Then, I will dissect the corpse. Therefore, it will eliminate the surplus that was not required during our time. Now, this appears to be more appealing. Therefore, we should proceed to the residence. We should conceal and select these items. Additionally. Okay, I will navigate to the view and toolpath tabs and display the axis. I like it. It appears to be flawless.

3D EXERCISE 3

We will commence with this foundation, which necessitates a rectangle with a length of 140, a width of 66, and fillets with a radius of 15 at each corner. Therefore, we will proceed to rectangular shapes and have 140 cross 66. A valid radius of 15 is also included. Exactly. I am required to select that particular point on the screen. Additionally. Then, the dimensions of 140 and 66. Additionally, the distance and the openings are now present. All right. The entire distance must be measured, as it is identical. Therefore, this distance is not specified. Therefore, we should establish it. There is minimal evidence. Therefore, I will draw a rectangle and select this point. Additionally. Almost conceal this. So I will create a 1540 at this time. I will then select solid extrude and then this one. Okay. How much extrusion is involved? So, oh.

The distance is sixteen kilometers. I will create a single direction and multiply it by 16 before reversing it. Therefore, we have completed this design. No problem. All right. Now, it is necessary to incorporate openings. Therefore. The solid that is being targeted is. Choose the desired target from the "okay" Therefore, there is only one. Therefore, it is not a short distance from that location to my link. I would like to include positions in this. Select this option. Select the "escape" button. Additionally, this one. And now, we will be receiving all of our relations.

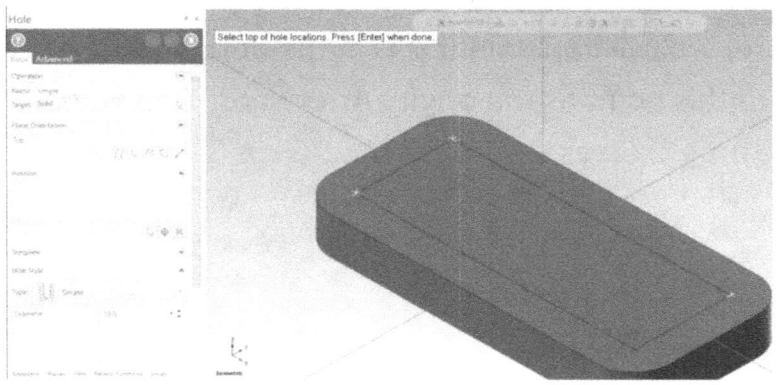

This is due to the fact that the other function employs points. Then, this one is implemented in the context of location. So, I will participate. I am now required to define the entire type. Therefore, it is countable. Essentially, the comfort mode consists of 15 outer layers and 19 inner layers. I will perform a modest diameter of nine, and this must be five. Additionally, the lengthy one is acceptable. If I. Despite the fact that he is. We are now required to complete this. Therefore, the most straightforward approach is to generate a design on the side and extrude it. Consequently, I incorporated the same item. Additionally. Afterward, I believe we should alter our course. Therefore, we will establish this entity. Additionally. All right. Therefore, I will proceed to. No problem. I will not locate it. Indeed. Additionally, I will be conducting a simulcast of this endeavor. And this one, and this one, and selection, and yes. I will now commence the process of generating. A rectangular shape. Therefore, we will implement this in the designs and designate this point as the origin. Afterward. The width will be. Three crosses seventy. 7381.

.

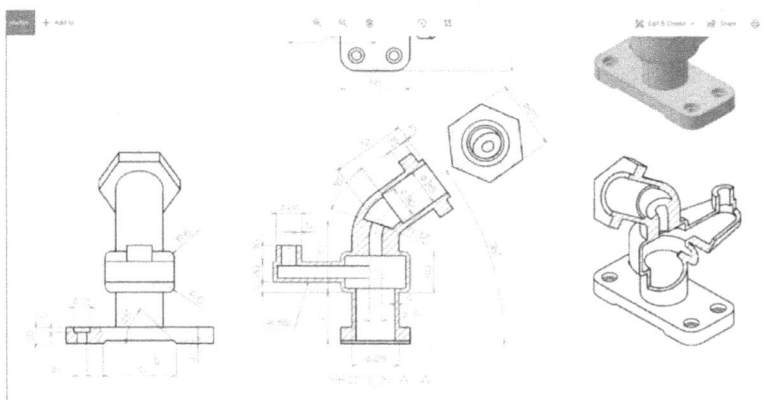

All right. Presently, the temperature is. 355. Therefore, I shall modify the configuration. This line and this phrase. Angle and distance. All right. 45 degrees and three faces. Afterward, this remark. The chain attachment is horizontal to the second line. The matter is satisfactory. Additionally, you are required to regulate your emotional responses. Yours. Therefore, the second line is parallel to the first line when modeling mode is employed. I am uncertain as to why it is providing a single value; however, we ultimately achieved the desired configuration. That is satisfactory. Navigate to the solid extrude option. Select this and cut the torso, then make two calls. Currently, we possess a file. Generate this symmetry. Our revolve command will be implemented. Therefore, I will construct this object within this geometry. Therefore, I relocate the master game for you. Initially, we will commence with an isometric perspective.

The following is an isometric view. Additionally, we will conduct psychological assessments. I will retain a portion of my perspective from this point forward. Additionally, I wish to conceal this from the front. Therefore, I will proceed to. Then, return to your residence. Keep hidden. Additionally, I will retain this identity for the sake of selection and distance. All right. I now have a friend on the planet of the functioning aircraft. Subsequently, I abandoned one. Therefore, the baseline is now located here. Initially, I will draw a single straight line that is 50 plus 7120 -30. Therefore, I will plot a line of 117. Therefore, this is the line in question. Perform the one hundred and one. Seventy. 117. Working them thereafter. In an isometric view, I will verify whether it is depicted on the companion plan or not. Therefore, it is evident that Franklin is not the location. Rather, it is situated within this context. Subsequently, I shall conclude that. Okay, I will proceed to the wireframe. Therefore, I will project this line and selection onto the plan for my frame endeavor. On which plane? Regarding Franklin. I apologize for this. I will be required to make a decision. An active plan for a specific aircraft. All right, I will proceed to my endeavor, specifically this line.

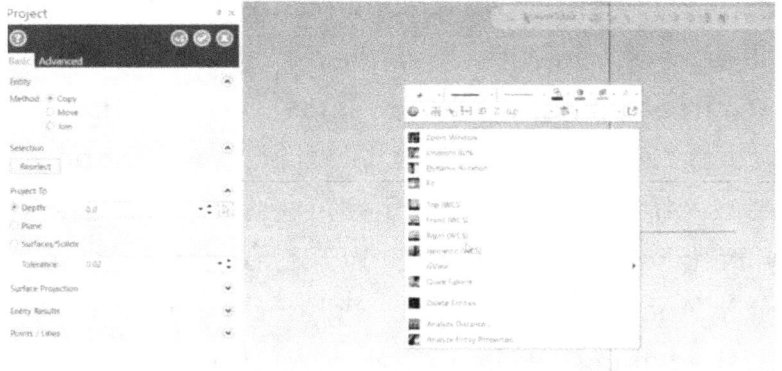

In reality, I will examine the denials and determine whether they are affirmative. Therefore, I desire to have my work projected. On this aircraft, everything is satisfactory. Here, it is projected onto this plane. I will complete this line and then move to the side. And I will simply retain this statement. Proceeding and. At present, I have aligned the stern plane with the view and will proceed to draw a line. 117 is the extent from this point. All right. Make it multiline. Also, another intermediate operation. Therefore, we have a location and a direction. Yes. I will begin the subsequent line at this point, as we are at the same position. Additionally. Yes. Therefore, I will not achieve verticality. The third line is from. Similar to this and scale. However, I will adjust its extent. It is currently 2 a.m., and the time is 70. At a 36-degree angle. Afterward, proceed to. All right. Distance. Additionally, ninety percent. 117. Exactly. That is acceptable; I will implement it. All right. point and marks to emphasize this line. I will provide you with this line and this point, and I will draw a freeform line and draw lines. This point is comprised of 72 and 36. All right. We have now limited ourselves to offsets. We should first assess the entity, and then I will analyze it. The number 117 is correct. Choose to accept it. And then there is 32. Right now. Sections will be established along this line. Therefore, the initial section will be located at 50, followed by 4090. Therefore, I will adjust the distance of this line accordingly. In what quantity? Fifty. Indeed, fifty. Fifty.

Subsequently, forty. Essentially, we must construct this section in this manner. All right. Therefore, the subsequent number is forty. One. and we are now obligated to. All right. It is imperative that we establish a boundary in this regard. The midpoint is employed. Therefore, the angle will be 90 degrees in addition to 3690 degrees. Additionally. Its length will be. I see. What is the amount? Thirty-five. On the left. Additionally, this line will be offset to the following: The dimension is not specified. And it is contingent upon this, this premise. Therefore, we should quantify it. Additionally, it is crucial. 65.695. Arrive. So, counterbalance this in the entity and this side, and control V. I mean, the fundamental structure is very. Beneficial for the. Exactly. Therefore. We will initially establish the outer section and subsequently offset it to the interior. Due to the fact that we have completed it in the "under" section. Additionally, the exterior region. Well, that's fine. Prior to. Therefore, this section, which is fundamentally. This is no longer necessary to select in these industries. Additionally, there is the line distance. Starting from this point. The distance for this line is 30. And we have a significant distance between us. Indeed, thirty. Therefore, this is unnecessary. So, as you can see, the distance from this point to this point is 50, and the outer diameter is 45. Therefore, I shall. Regrettably, I will simply offset the line. By means of the design. OK, I will violate it. Yes.

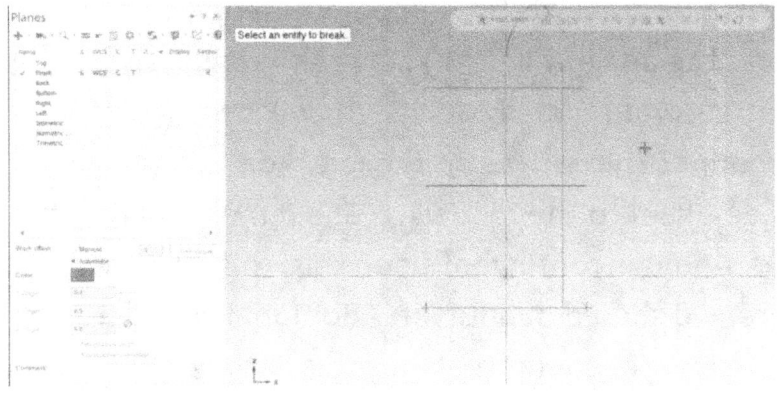

And this sentence. This is the one. This is the case. Subtract. This is the one. Additionally, we have projected this trajectory. In the same way as a bracket. At this time, we will eliminate this section. Additionally, this should be executed in an exceptionally pleasant manner. Clearly, this should be trimmed at this time. All right. Therefore, it is imperative that we execute this action. Additionally. Yes. You are aware of the issue. All right. Therefore, it is essentially the initial section. This is the extent of the diameter in the subsequent section. This point is essentially what I do, which is 66. Therefore, I will subtract two from 66, resulting in 33. Now, I will reduce this line, this extend, and this line on this line, and by this, I mean that every part. Therefore, you are required to produce it due to the fact that this pruning is so significant. Additionally, it will be necessary to establish a connection. Is the circumstance. Therefore, this issue is fundamentally interconnected. This is acceptable. And a steady line. This time, simply extend it. I am merely. Additionally, reduce this one. To this. Additionally, it is generating this line. Eliminate this. And indeed. All right. We have completed this task. And then there is entity. You. This distance is 40 meters. All right. In line with expectations. From an individual who is not within the age of forty. We have a time distance of 1 or 2 and an offset. You may assert that this is essentially the same distance as 45. Therefore, it implies that. Offset. Contains. Similar to this, this also moves. Afterward. Consequently, you are attentive. All right. I have completed the task. I am merely. Rotate this line. This is the case. Additionally, this geometry. I offset. All right. Additionally, what is the cost of this section? This section is available for your use. Be within this 235 and a maximum of 34.

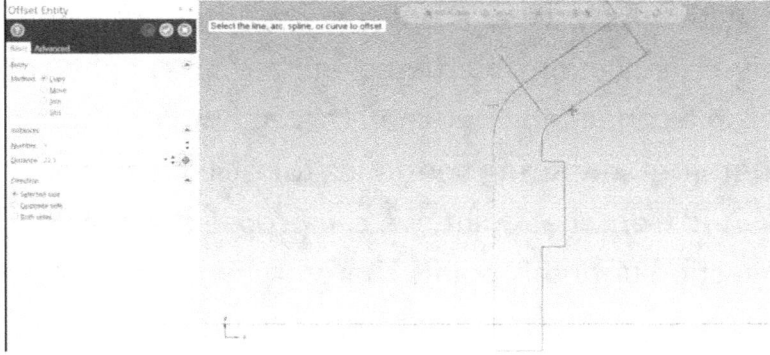

And also in this 35 by two. All right. And now, for this section, we will establish an offset of 15. Subtract it. Fifteen maneuvers. We will offset it by the same amount as it is now, which is twelve and six. To counteract. Subsequently, select this line six and this line, perform six. And it is somewhat in this vein, alright. So at this moment. This item is not assigned a name. Therefore, we will simply connect and delete it. Yes, it is twelve. And that is precisely what we accomplish. I will now draw a line from the center of this. Which will be associated with this point. Yes. This is in conjunction with the. I will now clear this section. We should move it. Therefore. All right. Rename this division. Therefore. First, reduce this line from -0. Trim this line and this will create this line. Additionally, I will eliminate this remark. Again, the center of these two is devoid of any content. I am now present. Similar to this line, this point, and this line here. Also, this one and the one in this line. Finally. I am establishing a modest section to serve as a single entity. Therefore, you will require this section and then eliminate it. All right. Let us establish a connection between this and the preceding topic. Therefore, my string continues to oscillate at this location. And from that point, we can proceed with a straight section until this point. Therefore, why not? We. If we have a reliable section up to this point. Therefore, we will require this identity. To this extent. Additionally.

Presently, it is flawless. Our sole obligation is to rotate this line. And we will establish a connection between these locations using an. In this paragraph. Up to this point. And this sentence. Up to this point. The subsequent phase. No problem. Now, we will employ the revolve command. Revolve. I apologize. It is now necessary to counteract this. The ID is the necessary motion, which is the upper portion. The position must be adjusted. Therefore, I will proceed to the wireframe. I will. Please refer to this line. And this sentence. All right. Afterward, the two were linked. Here, we are approaching the lower position. Therefore, our actions will be contained within. Positioning oneself at the lowest extent. Therefore, this is the extent of the situation.

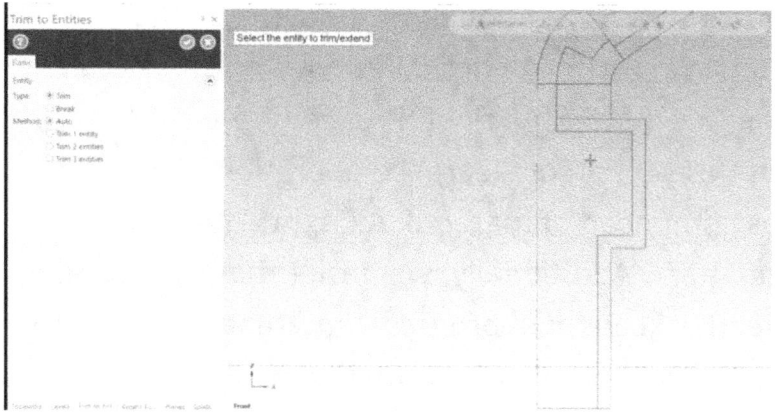

Therefore, we would shorten this line to this point. In that event, this line is unnecessary. Therefore, it is sufficient to modify this line. So it is one with "okay, back this line" and "this" and "delete." All right. Everything is now prepared from the top. Therefore, it is recommended that you opt for a robust revolve. All right. We must once more determine the boundaries. Therefore, we and we have an infinite radius of six. Exactly. Relocate the wireframe cylinder in a manner similar to this. Additionally, this one and radius six. And this one and this one are standing in the middle position. Exclusively. And indignation. It is possible to achieve this solely through offset. Immediately. Proceed to solid and along. All right. This sentence is unnecessary. Therefore, proceed to reduce this line. In this manner. Similar to. Additionally, it is imperative that we introduce this line at this juncture. In order to achieve it. Additional rotation. In this manner. Alteration. We. Exists. All right. We have both taken a stance on this matter. All right. Presently, we shall establish this section or clearance operation. The initial item was composed of material. All right. These two are presented here. Additionally, we desire that they adhere to this section. You must progress, and that is acceptable. Therefore, this alternative is primarily employed to maintain a level body. Therefore, it will not be perpendicular to the line. Therefore, I would like to demonstrate that unticked item to you. We will now incorporate an additional operation. Therefore, you will be responsible for this. I will proceed to. I am referring to. You are aware. All right. This section will involve the creation and movement of the aircraft. and an aircraft from a solid visage. Exactly. All right. I will now establish the appropriate values in this location. Combining the lines. and simply force it to provide you. Are you aware? The entirety of the information. This is the case. We are not interested. All right. Thirty. All right. I will now perform a solid sweep and sweep this circle. Additionally, this circle and this component. Additionally, it will. Afterward, we are required to participate in the forthcoming landmarks. Therefore, we possess this component. All right? We possess the necessary shape. Moving forward, we will pivot. This 30 and have this at this juncture in the previous. All right. Additionally, include specifics. Additionally, there is a selection. Therefore, you require this item, while we no longer require it. The error is being marked with a solid. Choose the identical option. Furthermore, choose it. Additionally. All right. Additionally, it revolving. This results in the correct rotation. This is followed by the subdivision view. I will also operate in this region. This will be the subject of the wireframe assignment, followed by the geometry. Surface on this site and section. Additionally, yes. Right now, I will return to my original outfit and continue to wear it. Is that correct? Additionally, I. For each. Then, is that illustration acceptable? And then, upon returning home, operate the mesh system with a single click. Yes. I have already made a mistake with one. Then, choose this organism and proceed to carve it. To appreciate. I recall that with a single touch, I will be home.

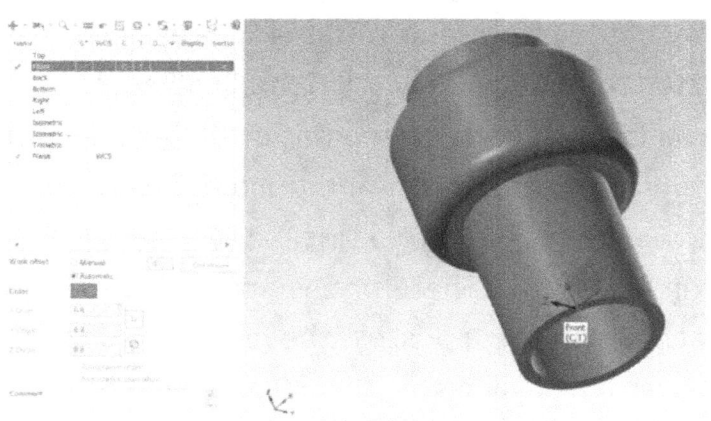

Yes. In terms of a metric. I will now proceed to develop this subsidiary section. It is imperative that we take action. From the incorrect to the correct frame. Allow us to. in diameter, 66 and. All right. Drawing a line of. And here it is: 82. From the center. Regarding. Additionally, eighty-two. Additionally, it is horizontal. In use. Additionally, verify this. Yes. Additionally, this. I will now commence the process of generating. Afterward, this occurred. I am capable. Import into my. All right. We will. Ascending. Right here. Additionally. If we have drawn a single line, we will receive it. Yes. Additionally, eighty-two.

Therefore, this item is included in the information. All right. All right. It is fifteen. Afterward. There is no rationale. Additionally, there is a border. Extremely exquisite. Engine. Acquire some and observe. Tangent is automatically selected as either in or between. Therefore, you are capable of producing. Information such as this, this, and this time. Just received. And to be this one and this one. sure, sure. Indeed. I will succeed if I work on the domestic side. Next, we attempt to determine whether everything is in order or if we should return home. No problem. Additionally, I am. Depart. Therefore, what must we accomplish in this location? Exactly. Break in the frame. Correct this and this point. This and this. And make this head this point and make this head this one. So, now, remove this outer section and. Dispose of this item and destroy this one. And at this juncture, nude will be present. Exactly. Additionally. Currently, the right frame project is being moved, and this is being done. It is anticipated that we will relocate. Therefore, we do not require this constant surface projection. Unfortunately, we are at a loss for alternatives. Proceed to solid extrude or. Yes. In both directions. And here is the quantity of this in 2010. "And in the event of mortality, because 20." All right. In this section, we acquire the ability to generate body checks and advanced modes in a single step. Exactly. Perfect. Create a facial expression at this time. To the oval. For the purpose of defense. So, this is where we must. We will generate these horizontal points or clicks in a manner that they are all connected. Exactly. We will relocate this aircraft and modify it accordingly. okay. Perfect. I will proceed to the appropriate frame and project from this point. On the surface, this appears to be legitimate. Of the and. I will now click on this and circle it. The central item will be acquired. Here, I will generate a profile diameter. No problem. Additionally, it is twenty-four. All right. Additionally. Select "offset." I am intrigued by this. There are five of these. Systems and. Exactly. Foreign mode. All right. An error has occurred. Subsequently, increase the number to five. Please specify the current status. Presently, extrude and solidify. Regarding it. Is this distance 16? 16. Yes, I see. Every time, ensure that we have a ten-selected item. All right. Right here. Once more, for. Additionally. Cut this by clicking on it. Active. Already. Exactly. Also, modify the direction to indicate the amount. All right. Because it is twenty, sixteen is acceptable. Additionally. I will simply prepare it, as we are only interested in preparing this particular portion. All right. Alternatively, you may select this headgear and remove it prior to use. All right. I will now proceed to the exterior. Shell command. Essentially, the interior body is the next step in the body incision. Let us simply select this visage. I will enter the face and select this body. To maintain the entirety of my thinness on this body. We are opposed to its inclusion in any dramatic scene. Well, that's fine. What have I accomplished in this section? You. We should prioritize it. Yes. Section. Therefore. What is visible within that indicates that everything is in order? The issue is simply that. Additionally, this form. All right. Therefore, I will proceed with the operation. This extrusion is the reason. All right. We merely wish to extrude the exterior. And bring this extrusion. Please proceed to the subsequent shell.

Therefore, it is evident that the sequence has been altered. Therefore. We achieved this by altering the sequence. Presently, the sole obligation is to eliminate this region. Consequently, I will be required to visit the picture department for this or another reason. And indeed, we must initiate this. However, we once again overlooked that identity. A demonstration. Indeed. That is not the case. All right. All right. We made an effort. I will now. Complete all other tasks. It is necessary to remove the portion. I must navigate to the wireframe and offset. Exactly. Conduct the project from the beginning and focus on the chain. Therefore. You are aware that the thickness is the. All right. This thickness is acceptable. This serves as the sole objective. It denotes a star. Therefore, remove it from the indicator that the form should be acceptable. Afterward. At this juncture, we must trim the portion to construct a two-point engine circle for the circle. The subsequent premise is as follows. The typical initial site of incidence. This location and station. Additionally, I will adjust the radius to 39. 66 cans. Therefore, the header was added at the conclusion. No problem. Now, we must eliminate this section, which is essentially added to display at the beginning. Subsequently, preceding this one. At this point, we must eliminate this section. And only this portion, the portion that is between this line in this name. So, I will simply eliminate the remaining items. Also, I will now advance this point. Additionally, from this location and this one. All right. We do not require this. Introduce a new delete and. All right, that is satisfactory. All right. You possess this item. Therefore, we should return to our residence. Keep hidden. It is imperative that we do so. Select the inner. Intersection or omit the wife's name to enable us to view the right side from within the intersection. Therefore. I now proceed to the extrusion of the legs, and I am attracted to this and the same. Yes. Select both directions. And I will make it eight because we have a distance of 14, which is four out of four in laundry. Four plus four is eight, so we will have 25 and make it six on one side. I then proceeded to remove the cadaver at the designated location. Solidity is the case. This is the one. Additionally, the notification does not activate game direction. Once more, I opt for the same option. Additionally. Additionally, this time, I opt for this sturdy. Consult the correspondence contents. All right. I will now return to my residence and conceal myself. Currently, this section appears to be in excellent condition. Exactly. Yes, the interior portion is now vacant, indicating that everything is connected from the inside. All right. Therefore, I apologize. The final operation that we must perform is satisfactory. This region will serve as residence. I possess none. Where is it necessary to construct this structure with a diameter of 67 and an incorrect thickness? How dense is this? The thickness is fifteen millimeters. Additionally, it is an offset that is five offsets from the horizon. To construct an aircraft from 100 stars. Additionally, this is your visage. Certainly, allow me to demonstrate one item. Additionally. You simply wish to rotate slightly, correct? Yes. Presently, this aircraft is the sole one. In the event that I select the aircraft, it will display the airplane three six. Subsequently, this is the primary characteristic of plane one. Yes. Additionally, we will determine its z value as or 146 minus five. And it is currently in the ideal position or acceptable. It is consistent with or. Yes. We are currently in the process of reviewing this message. Therefore, I will proceed to engage in the act of writing. The same rectangular shapes, polygons, and yes, six sides. Then, to the main topic. Yes. And so forth. The radius is 67 meters. I will verify the value of 67. I would like to. It is 67. All right. At forty, I abruptly realized that everything was right. We should verify this. Therefore, it is contained within this. At this time, I apologize. Outside of this, everything is fine. Therefore, click on this polygon. This is the same idea. Additionally. The number 67 is divided by two. We will also make a phone call. This time, it is acceptable. Additionally. The circle is located within this area. It is exceedingly level. Line. Proceed to the secure destination. Form I. Segmentation is satisfactory. And construct a body, rather than relying on images. And what is the amount? Yes. 15. Three. Fifteen kilometers. All right. All right. Unfortunately, I neglected to include an essential item. Therefore, I will proceed to the broadcast and endeavor. We must project this in this instance, as otherwise we will be unable to do so. Alternatively, we must conduct the task in a different location. It is possible for us to return home. All right. Proceed. First, extrude and select. Yes. This is the one. Afterward, this will occur. This will be the result of the creation. Afterward. One side. All right. Home is no longer present. All right. Additionally, I mentioned this location. Also, navigate to the "view" option. Conceal it. All right. Then, I am symmetric and accessible. All right. I will now activate the second section. Viewing is feasible. I have no cause for concern. Is it true that everything is void in this location? The drawing has been finalized.

SURFACES

Surface is our most recent subject. This is a distinct entity from a solid, as it has no thickness and is frequently employed to represent objects that are too intricate to be accurately represented using k-means. Therefore, to examine the outer edge of any board. Therefore, we employ the term "surface." I will reach the top if we proceed to examine and on this. The most engaging individuals. And now, I will deactivate the switch and ascend to the surface. Additionally, I will construct a fundamental cylinder. Additionally, if I activate section view and select isometric, the following will occur. It is evident that there is nothing. Additionally, the term is pronounced. I am required to create the front section. Yes. It is evident that the thickness is negligible. Only the exterior portion. All right. Therefore, it is equivalent to firm for all commands. Also, if I generate a blob, the same thing will occur. However, the thickness is negligible. In a similar manner, the sphere is composed of an outer layer and a porous final layer. Yes. In that location, it acquires its form.

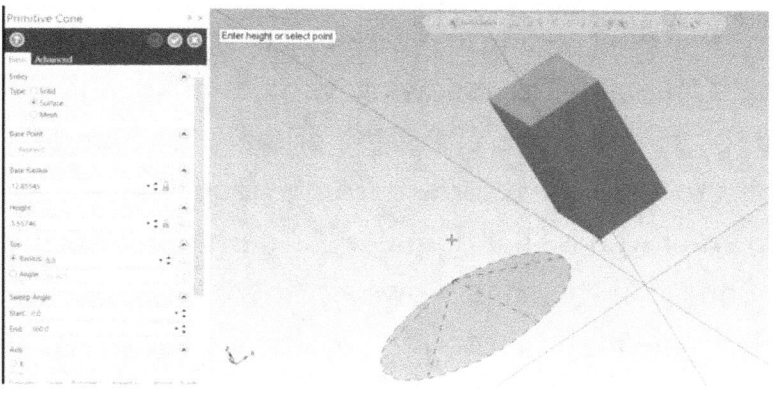

Afterward, you will be able to observe the entirety of the section. There is no substance. Therefore, we should delete the item and set it to inactive. I will now establish a solid. In the same way, I will construct a cylinder. In the event that I wish to convert it to. If I wish to convert it to a surface, I will navigate to the "surfaces" section. Solid surfaces. Please select this link. And here, we have a choice. Click three times to select the body, and then press the shift key and click "OK." Therefore, I will eliminate the number 123. Now, each word in the section has been selected. And we will depart. The complexity will be increased as a result of the division of these parameters. Now, it is evident that it is designed to be visible. However, if I proceed to solid and. Will disclose. Additionally, this location. All right. If I eliminate the solid. The material from which the surface is constructed is evident. It. All right. Next, we will eliminate this item. Also, from aircraft. I will be able to observe it from a distance. I will now generate a wireframe. Let us establish a rectangle. All right. The following is a command. It is referred to as a folio. Therefore, the primary function of the page is to generate a surface from a good. Therefore, the term employed is "flat boundary." Therefore, mobility is a component of the page. However, there is no issue. A surface is visible in this location. I would now like to demonstrate another command. Similarly, we have used solids to cover the elevation. Therefore, we will examine the exterior of our loft. Therefore, I will establish a land that is proportional to the units. All right. At the very top. The quantity of offset that we desire. Let us complete the entire collection of thirty. We should aim for sixty. I will now establish a circle in this region. Additionally, the radius is 25. I will now convert the design to a surface by employing a planar boundary. Additionally, that is acceptable. Transform it into a surface. Afterward. I will include an additional plane in relation to envelopes. I require an offset of 30 in this plane, and I will proceed accordingly. I will now generate a polygon in this location.

Not ellipsis I. In locations where its radius is considerably extensive. okay. To specify. I will convert it to a surface and then proceed. I will now employ three complete, planar planes. So what it will do is, and I apologize, three surface planes will be filled. Therefore, it will deform three surfaces to generate a new surface. First and foremost, I must proceed in this direction. Everything is directed in this direction. Secondly, it will be in this direction at any given point. Therefore, it will be necessary for me to modify the direction and timing. Is this the one. Well, that's fine. Also, submit. Therefore, we will investigate the formation of this peculiar geometry. So, I will adjust its magnitude to my liking.

You are aware of the concept of shape change. In the same way, the shape of my 2D transforms utterly when I convert it to one, one, and under magnitude. Therefore, I desire to be permitted. Therefore, it is employed for this purpose; consequently, I shall terminate it. I will now employ the "trim to plane" function. Therefore, I will select all three and six and utilize the aircraft that is available to me. The first element is the line, and the second element is the cursor coordinates of 3R2. Therefore, I suggest that you commence with the prescribed criteria. Therefore, this is the initial argument. The second is this one. Additionally, this one. Therefore, this is the method by which a plane is generated. That is entirely different. Therefore, we should proceed. Subsequently, how can one locate it? Yes. Next, I will generate an additional aircraft at this moment. All right. I will succeed in reaching isolation. Therefore, it is evident that this section is similarly trimmed. Then, we can create a controller. In the same way, we can trim two surfaces. Therefore, we will refrain from delving into the specifics of the surfaces, as they are not utilized in the machining process. Therefore, I will proceed to the wireframe and begin to sketch. I will construct a glass-like object with the assistance of the horizon. All right. Initially, we will proceed to the stand aircraft. Yes. The aircraft belongs to my acquaintance. I will establish an organism that is similar to another. All right. Additionally. Afterward. I will now ascend to the surface. I will travel to the moon and choose this aircraft. Additionally, the axis of rotation and the angle. Regrettably, I neglected to relocate the line wireframe. Again, navigate to Surface and Revolve. So, let's enter and observe that we have already implemented Hello World within. All right. Let us assume that we have been granted dominion over Z. Now, if I wish to construct a conduit similar to this or something else. Therefore, we should proceed to. Yes. We should conduct a cleaning. It is possible that it will be located at the bottom. It is essential. Additionally, I am interested in developing an object on this aircraft. Therefore, I ascend to a higher frame and circle. I would like it to be perpendicular to this. Therefore, I will transfer to an alternative aircraft. Exactly. Plane. All right. I will now establish a circle in this location. "And I will simply possess this." Way to expedite this process. Additionally, this section is acceptable. Therefore, a pipeline has been established. In the same vein, the majority of the command is also secure.

Therefore, we should not squander our time in this line. Additionally, we will proceed to a new section titled "mesh." Additionally, mesh is not employed in machining; however, it is employed in 3D printing. This is the fundamental distinction between the lattice and the solid. I will demonstrate this by creating it. Of course, there was a default setting when I created this. and now, if you can observe, when we construct a solid. Yes, there is something upright about it; however, mesh mesh divides the body into small features. It is primarily employed in simulation analysis and calculations, as it divides the area into small faces, such as a rectangle or polygon. All right.

Therefore, meshes are generated from entities. We have the ability to generate geometry from the solid. In the same vein, these are all of the commands.

MATCHINING

We will now commence our primary subject, which is machining. Therefore, we should commence with a fundamental model for the purpose of machining. Therefore, I will proceed to the wireframe and generate a diamond-shaped box. Okay, let's relocate that aspect to the center. Additionally, it is one hundred. Then, on the Android Cross hundred. No problem. No problem. We should extrude it into our. Unfurling. We should establish a circle within the area. 60 meters are comprised of nine meters. We will simply conduct a thirty-minute meeting. All right. Not thirty. Complete the task. Food that is either round or substantial. It. To be extruded. Exactly. I will now perform isometric calculations. All right. Therefore, I will initially proceed to the state machine for machining. And here, we have alternatives. Therefore, we will commence with the mill. In this course, we are currently in the process of wiring a mill that we will refer to as such. Additionally, these are examples of sophisticated machining. Therefore, this is a list of devices that has been defined. Therefore, we should determine whether it is initially designated as the default designation. Therefore, it is imperative that this be recommended. So, this gadget is primarily utilized for metric measurements. Additionally, this is employed in the measurement of inches. Exactly. Therefore, we typically employ the default computer.

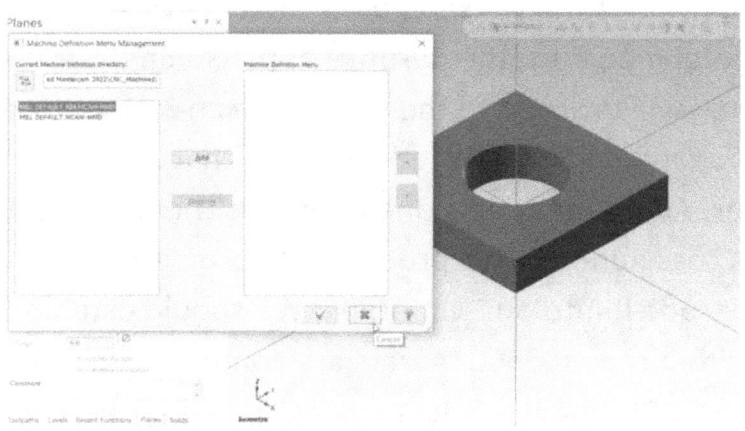

Additionally, we will incorporate any machines that we possess into the library. All right. Therefore, I will return to the default. We will now proceed to the instrument. I have reached the conclusion that I would like to return to the default setting. Then, properties are mentioned. Therefore, it is imperative that we verify one item prior to the creation of stock. Therefore, Dublin is the intended instrument plan. Therefore, we will once more proceed to the board and initiate the preparation process. I will now establish a bounding frame. We have the option to specify an alternative type. Additionally, it is evident that the stop line is in the process of crumbling. Therefore, this is a specific form of bounding box. We have the ability to generate cylindrical and rectangular shapes, but we must specify a single axis. This is a solid lattice, and the axis is located in that area. This will be the subject of our discussion. We have the ability to select any other solid that we have created within this triangle. It has been imported from an alternative file. And here is where we can choose an external file system. Therefore, we will commence with these two. I will not address this issue in this project. because it will be somewhat challenging. First, we will cover the area, and then we will proceed to the bounding box and control a to ensure that all items are selected. Additionally, there is the alternative. Therefore, the minimum fixed size with no excess stock is 100 120.

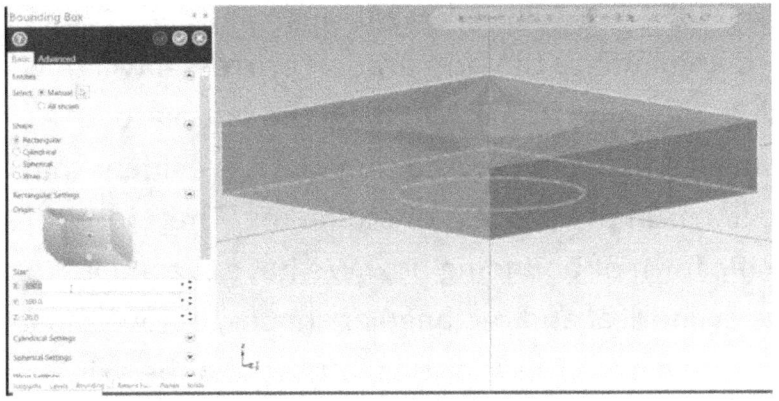

We will now increase the number to 110. Additionally, 110 and 25. Okay. This dog can now be transformed into a solid. Upon clicking on this, it will become firm. Therefore, our substantial file will be generated. Therefore, in this instance, we will not generate a solid, which is why it is marked as unticked. Additionally, that is acceptable. All right. Therefore, its current origin is zero zero and 22.5. So this value is essentially its source. Now, we select "OK." I am currently discontinuing the use of the stock configuration in the model. Indeed. Therefore, this is the best that you can observe. Suppose that we desire to avoid any stops below this point. Therefore, our course of action will be as follows. We will return to the stock configuration and bounding box. Bounding rectangle. Once more, we desire to select control A by clicking on it. And now, we will create the origin of this portion. I now mean that there were none, and there were 110 and 25.

Now you can observe that there is no stock on the ground if I were to move from this location to the right. All right. I am currently on the travels. As a result, I saw it. Yes. As a metric. For the time being, we have only generated stock on the upper side. Additionally, we do not possess any inventory at the base. All right. So an additional option that is available on this auto orientation that we will already be using in the face if we wish to align our stock with any face. Therefore, we will select the specified visage by selecting on it. Therefore, we will proceed with this option for the time being. Afterward. A stock is updated. Initially, we will commence with the most fundamental operation, contour. Therefore, we should simply select cryptocurrency. Therefore, what modifications are we proposing to implement? So, the control system essentially follows a part chain, and the drill bit, like any other instrument, will rotate and perform a cutting operation.

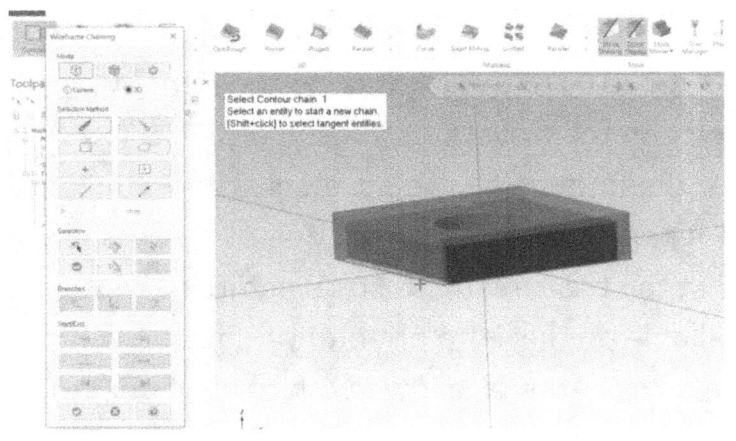

Therefore, it is primarily employed for the purpose of executing outside patrol and outside band. All right. Therefore, I will simply select this option. And we now have additional alternatives. There are several libraries from which we can generate a select tool, so operation is the control right tool. Therefore, I will opt for the Tool Library. Many instruments are now available. Therefore, we are unaware of the specific instrument. This library is a collection of masculine instruments that are essential for the well-being of male members. OK, we can establish our own library. Therefore, for the time being, we will continue to employ the standard tools, and I will implement the filter by selecting "filter." It is all in, and I will take one month. And from this point forward, alright. Everything is in order, gentlemen. We are not currently in need of any specific material. Therefore, we will implement the flatten main method. Therefore, this constitutes the initial component. Therefore, the diameter of the instrument is a radius.

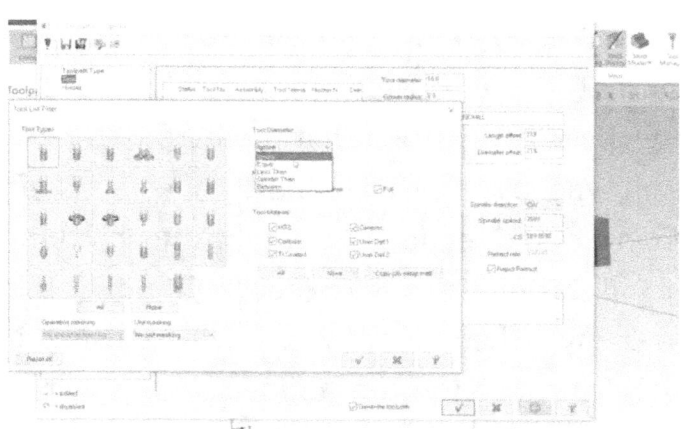

Let us assume that the desired value is five. Additionally, I will implement the CV filter in this location. This instrument. All right. Presently, verify the inventory. What is the length of the tool? Therefore, the duration is the issue. Along its entire length. Additionally, we possess a length of 20. Therefore, we will now alter this instrument. Instrument for editing. Given these circumstances. okay. Additionally, the trimming length is eight. Therefore, we will establish a cutting length of 25. Therefore, proceed to the subsequent step. I am aware that this instrument will not be implemented. In reality, they exist. However, for the time being, we will simply establish a cutting length of 25. Next. Complete the task. Exactly. Tool selection has been completed. We will now investigate the circumstances surrounding the planning of this. Observe the tool removing the likeness and conducting an incision on this object. If the instrument is visible, it is located within the material. Therefore, we will proceed to the third parameter that we possess. Additionally, you. Therefore, we have established that the barometer is a compensation type layer with a default value and a left-facing compensation direction. Therefore, the instrument is proceeding in this manner. If I select "right now," it is displayed as "okay." The point at which we are using tip compensation for cut center is as follows. In this instance, the middle has been flattened. Therefore, whether we choose any option, it will not have any impact. Therefore, it is advisable to abandon it and begin to remove it at a later time. Therefore, if I select one, this is the quantity of material that will be removed. Therefore, I am adamant about not leaving any material behind. Similarly, it is placed on the floor and it is determined that these diameters are unacceptable. Subsequently, we shall proceed to the use of depth incisions.

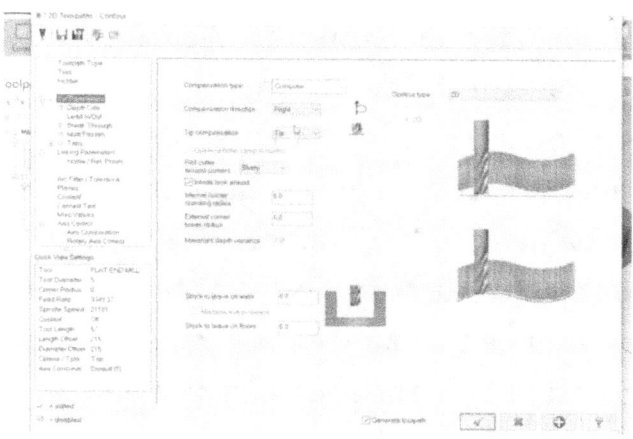

Therefore, the number of cuts required is 25. It is impossible to cut all 25 films in a single card. Therefore, we require a few IDs. Therefore, this is where we divide. Therefore, the roughest stage is the most important to him. Therefore, we should limit the duration to five minutes. Devoid of sensation. Therefore, these are the final edits. We require only one final trim. Additionally, what will be the step size during the final cut? It will function satisfactorily. Spindle speed and feed rate. These are acceptable by default. step. We will not engage in discussions regarding these advanced options and will instead proceed to the breakthrough. Therefore, breakthrough is the extent to which the material surpasses the bottom of the container. Therefore, if I select the drag tool, the value will be 0.01. All right. "Yes." Right now, Marlin should relocate. Performs multiple tasks simultaneously. Depth cut illustrates the distinction between multi passes and depth cuts. We have designated permits for the depth. Therefore, the instrument is currently in motion to the side. Therefore, if the side distance is excessive, it is impossible to operate the machine one at a time. Therefore, we will implement multiple steps for this purpose. I will activate multiple passes and determine the number of passes required. I would like to ensure that one roughing pass and one concluding pass are included.

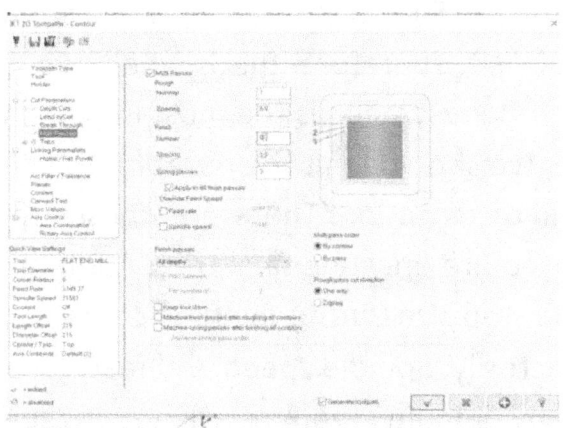

Therefore, the roughing in spacing is five. And for the finishing spacing, 0.5 is acceptable. Therefore, if I deselect this option, it will apply to all concluding passes. The spring passes are utilized to remove material; therefore, we will refrain from employing them and instead divide and conclude with these diameters at all depths.

Additionally, it implies that we do not wish to have, say, five depths. Therefore, we do not require a concluding pass for each depth. Therefore, we will only achieve the final depth. The machine concludes after roughing all current machines, and the spring passes after roughing all endurance. Therefore, we will not implement any of these alternatives at this time. That is the meaning. In this scenario, the initial machining process will involve a depth cut, followed by the dormant depth cutting process. Subsequently, the machining process will be completed. Therefore, we will not delve into it in detail at this time and instead leave it in the coupling parameters. All right. One item that we occasionally employ is a profundity of five and a maximum stock of 25. Therefore, if I do not wish to remove the entire portion, I will specify a depth. Let us assume that this is a point that we wish to eliminate. At present, we intend to terminate the project at this juncture. Additionally, it is zero. Therefore, the profundity is zero, which is acceptable. The subsequent step is to simply select "okay" and exit the page. You can now observe that there are multiple passages.

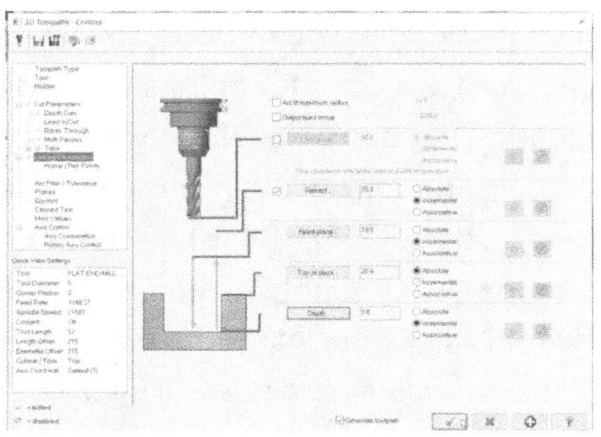

It is the initial depth incision. The third depth cut is the second level cut. Exactly. We are currently in the process of determining the final pass. As you can observe, it is the final pass. All right. We will now proceed to the subsequent machining operation, which is the drill. I will initially proceed to the solids in this location to begin drilling. Additionally, the wireframe and the focal point. Therefore, this is the case. All right. I am required to return to the given direction. I will now generate openings at random locations. All right. I am uncertain as to whether the diameter of ten is accurate. To create this one with a diameter. Ten. In the same vein, this or. I apologize. Additionally. Then this number is from 1 to 10. All right. I will now extrude this curved shape. Extrude. From this point, cardboard is used to create one. All right. Additionally, I will extrude this center. Develop polymers with only five components. Subsequently, it is acceptable.

For. Our model has been altered. Therefore, we will proceed with the drilling operation.

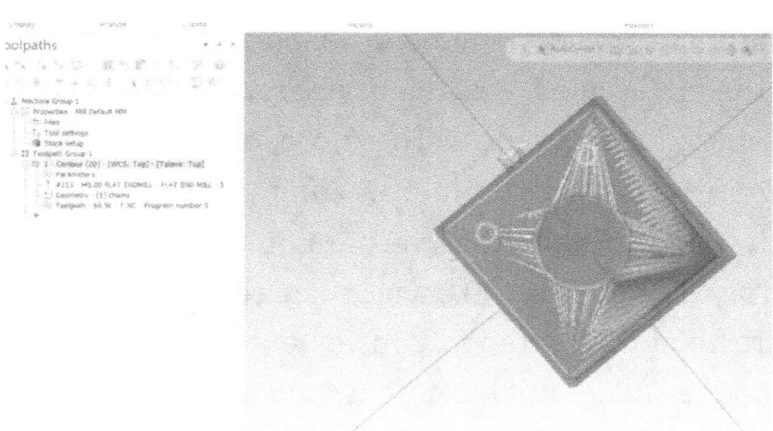

Therefore, proceed to the mill tool inspection. It is time to let go. Therefore, it is customary to execute an operation card prior to drilling. We should relocate it. Additionally, the same. Drilling at specific locations. Consequently, spot drilling essentially generates a limited area for the drill bit. Therefore, it will not be able to travel. Therefore. All right. False. Click "OK." I will proceed to the instrument from this point. I will once more visit the critical library group. There is none. Tim's stepper parameter and radius will be located within this body. If we hover over it, it is more than ten minutes at this point. Eight hundred beehives. The administrator or someone similar will be responsible. Espadrille is the item in question.

Additionally, select it and select "diameter." And meter. It will return ten. Therefore, this is the situation. Therefore, select "OK." Additionally. What was the value? Yes. This section is satisfactory. No variation is generated. False. It is evident that it will not enter the building. It has recently revised the minuscule I. Yes, that is acceptable. We will now establish its parameters. Therefore, proceed to the parameter that is to be trimmed. A comfortable drill is also necessary to ensure that the parameters are satisfactory. The depth is "0", which is acceptable. We are not interested in this scraper. Consequently, we intend to both produce and identify. Let us assume that the stream is simply being trimmed. Therefore, I will include it in the compensation, as it is acceptable. Therefore, disable compensation for temperature dips. No, we will not generate or discard anything. Therefore, we should verify that we wish to blade. The instrument is not even moving to the surface, as is evident.

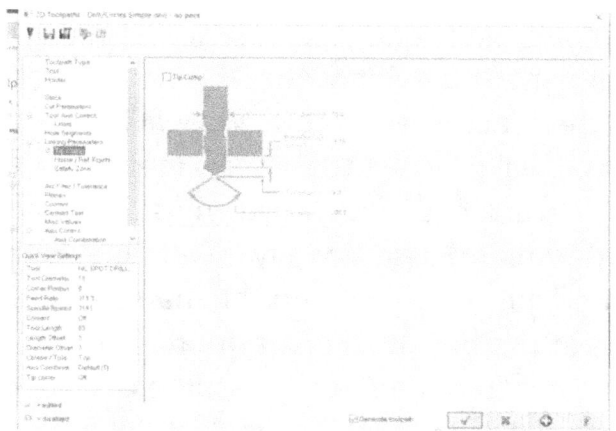

The reason for this is that we have increased the depth to three. Therefore, this is the zero, and three is upward. Therefore, it is imperative that we achieve a temperature of -30. Therefore, I will navigate to the linking parameters and modify the value to minus three and into. I am aware that this will only lead to the. We are currently in the process of incorporating an additional drilling operation. And this time, we will choose the and yes. All right. This is where we can modify the beaker's tip and bottom. Similarly, we, the commanders, should be permitted to proceed. Therefore, this utility is of no consequence at this time; we will return to the Scilab tool library filter. Additionally, we will employ a compact drill to drill this time. Produce. Yes, this one is already satisfactory. Equivalent to ten. One option is to reduce the amount of carbide. And then there is the adjustment period. Therefore, we are not concerned with the materials for future use, as the length of this one is 49 and the current one is 71, which is excessive. All right. Therefore, select

"OK." Additionally, we should evaluate what has transpired. All right. It is evident that it is not being inserted into the reactor. Therefore, we will now establish its parameters. I will proceed to the parameter that is to be removed. Therefore, do we desire any additional leisure to contemplate? Additionally, dwell time is the duration of time during which. And the material tool simply rotates at the bottom. Furthermore, for the duration of the allotted time, let us assume one second. All right. Additionally, navigate to the duct linkage parameter and specify the depth. So, I will select this function for this purpose. All right. Additionally, hit on that. I will now select a border. Proximity. Additionally. Okay, I will choose any of these. It is currently -20 degrees Celsius. And now, typically, recompense. Then, insert the v-one into the instrument to eliminate it. Therefore, let us assume that 0.25 is acceptable. These will be the outcome. Therefore, 0.25 and then downward. Moving this will be satisfactory. And estimate, alright, I see the metric is if I and approve it. It is. Going to be a condition. Yes. It is evident that the instrument is protruding slightly from the base. And that is the apparent victory amount. Exactly. All right. We will now proceed to the new operation's front. In general, the initial operation is to face. Therefore, we will be compelled to select this outer chain, as there is no further option.

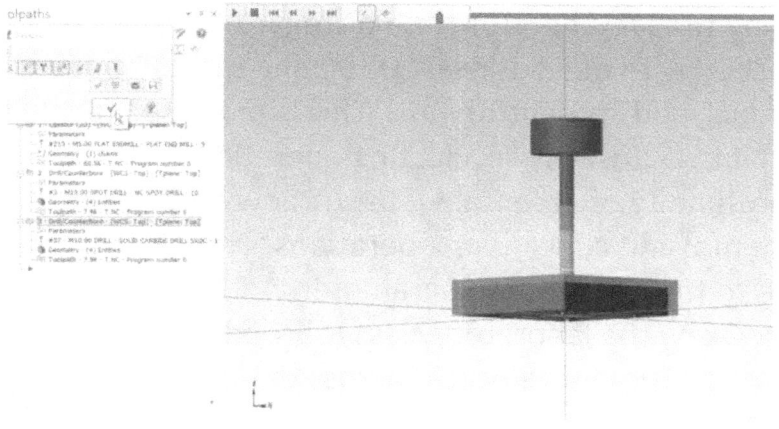

Therefore, I will switch to 3D in this instance, as it is merely a wireframe. Presently, in three dimensions. Yes. Furthermore, choose it. I will be selected in this manner and will specify the instrument I prefer to use. And here it is, with its face toward the viewer and its diameter. Let us assume that it is equivalent to ten. Therefore, we will generate this location due to our lack of resources. Therefore, you will select "First, determine the cutting diameter." achieve it. We should convert it to a secondary diameter that is three-dimensional. Twenty-five. Afterward, the total duration will be determined. Thirty inches of cutting length with five acceptable and ten for a 45-degree shoulder. Additionally, increase the number to ten and proceed. All right. It is now time for them to follow. The length offset limit is as follows. Additionally, the primary quantity of material and the cutting pace in the surface. These are the parameters that have been provided in the form of a table for each material and type of machine. At present, there is no specific material or something similar. Therefore, we disregard this matter and proceed with our business. Obtaining the component will be quick and straightforward. This is the case. At present, it is exclusively conducting operations at the single profundity of acceptable. Let us now define the experimental and concluding operations. Therefore, navigate to the appropriate parameter. Additionally, this location. Okay, the laboratory class is based on the rate at which the instrument relocates across the body. And along the limb of our laboratory, which is this one, the quantity is the distance that the instrument travels along the body. All right. In order to specify the specified system, we must traverse this. Therefore, it is evident that it is situated in this location. We should eliminate it. This one is a C. This is the one that is along. Additionally, this one poses an issue. No changes were made to the original sentence. And here, if I allow it to become slightly tighter. Therefore, this will be a cross when this neuron's instrument is released. Therefore, navigate to the parameter and observe that it is a cross with a 10% value. Therefore, 10% is nearly there. This is the reason why the instrument portion is visible next to it. Therefore, the stock should be left on the ground floor. Additionally, there are depth incisions. Therefore, the utmost number of roughing steps is five. And let us achieve it. Indeed, I apologize. The quantity of concluding incisions. The initial line and the final step will be combined. And then, alright, leave the default twinkling parameters at the top with a depth of zero. However, we can simply define it as zero. The inventory's maximum quantity is twenty-five. I believe it is currently being executed in accordance with our definition of operation. Hallelujah.

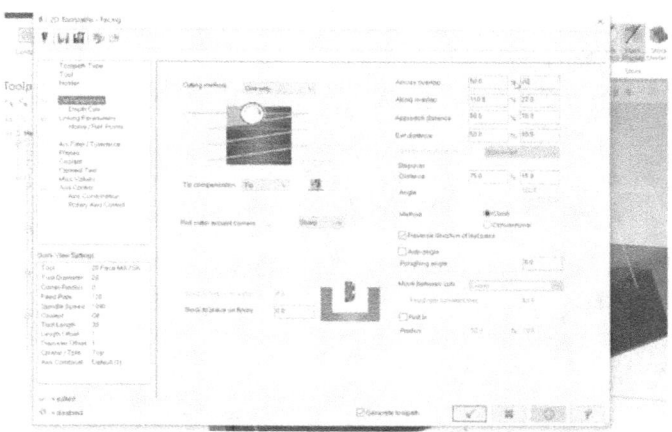

Okay, let's modify the crossing. The crossover lap is 10%, so it should be 20. Therefore, you will be able to identify it. When it was ten, there were no operations and no 2D elements. However, when I increased it to 20, they were present. Another advantageous feature has been incorporated. Okay, it is a cross lot. We can move this operation to the appropriate position. Therefore, the initial operation is typically to face. Let us proceed by confronting it. The initial approval is granted, and two additional approvals are granted. The initial approval permits the execution of the second drill and the vulnerable drill. It is a location that is acceptable. What we will now do is employ a compartment. Presently, we are. You are alright. And for the wallet. This is the chain that we employ. In the purse. Alternatively, this type of tool component is designed to be stored in a purse. Navigate to the "Tool" menu and select the flat in the center. In this manner. And the parameter that is cut. No single stroke is permissible. Additionally, roughing. Percentage or step

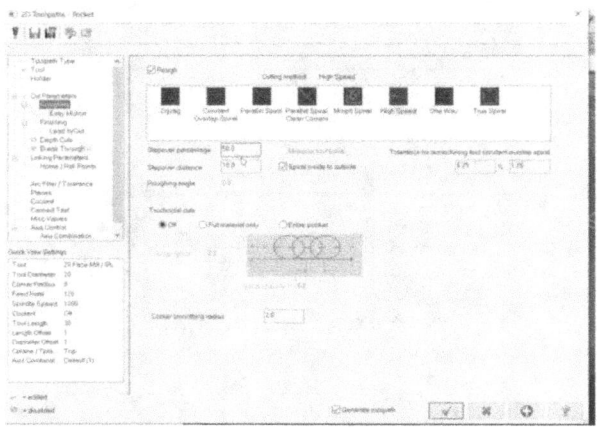

Cross the distance and allow for a spiral to form within. Therefore, these are a few of the components that are currently being utilized in a spiral fashion. This is the one. Therefore, if we pass over the means, the distance between these two, three components. Additionally, it is typically equivalent to the tool's diameter. Therefore, the entrance motion and the conclusion. Therefore, the interval between phases is 2.5. We do not have any spring paths, and we begin the process of finalizing paths at the nearest location. Indeed, the outer boundary has been satisfactorily completed. These operations and local depth cards will not be implemented. Therefore, the utmost number of roughing steps is five, with a final finishing step of four. Breakthrough linkage parameters and depth are now available. We have. We selected that chain for a reason. Consequently, it is classified as in depth zero. Begin at the top of the page with 25 and select "OK." Right now, I am. Observe the spiraling motion. All right. Machining has been finalized for this component and this component. I will now select the machining group one and two machines, which are minute. Presently, we shall verify the state of the tree. Additionally, what will be the machine's conduct? Therefore, this is the case. Within this organization. It is a massive system. Therefore, we will deactivate this equipment and its enclosure. I will now proceed to view and adjust this screen in this location. This topmost portion of the instrument is no longer required. Therefore, I will conceal it by employing tool components. One layer of tendon is sufficient. Therefore, the instrument encompasses all present operations in this and this segment. In the same way, trace and other trace refer to the toolpath of the upcoming tool and the follow-me function, which we have successfully implemented. Additionally, this is the initial stock. Therefore, initial investment is unnecessary. We do not require six. Is my preferred stock and instrument acceptable? They are currently in the process of removing it. And here, we will be able to decrease the simulation pace. And there is a spectacle. Therefore, it is contingent upon the system. All right. Speed of play is 200, and it is necessary. Decreased. Yes, I will encourage you to make a large splash. How are you doing? Also, you proceed to the beginning. I am now going to perform the initial roughing operation after confronting. And then the conclusion. Exactly. The control left executes edits. Additionally, to conclude this section. drilling. drilling. You are aware, in the pocket. Observe, we have made a significant error. I will now close it. I will then put this set of meters in my purse. We now have a pocket that is beneficial when you access this utility. It is imperative that we utilize this apartment. Contained within my grasp. Face me has been implemented. All right. Once more. All right. Proceed to the machine once more. Go to the simulation. You know, if I. Yes. The timetable. Operation is demonstrated. The initial operation that involves the intake of. I am not. Whom? I am. drilling. Presently, it is functioning satisfactorily. I am the ideal component. The rapport in which this can be said is another feature. All right. Even in its movement, the instrument is a work in progress. All right. If there is any collision, we can disclose it here. Now, if we wish to save this model, we have the option to navigate to the simulation and save the stock. Therefore, I would state it here, Mr. Our stock is now available in steel file format.

Right here. You navigate to the 3D view. Therefore, it is a substantial component. Therefore, in certain instances, the machining operation is performed using a distinct machine. Groups refer to distinct sequences within a single cycle. Subsequently, we generate and archive. This stock remains in use for the subsequent machining group or sequence. The most recent machining operation.

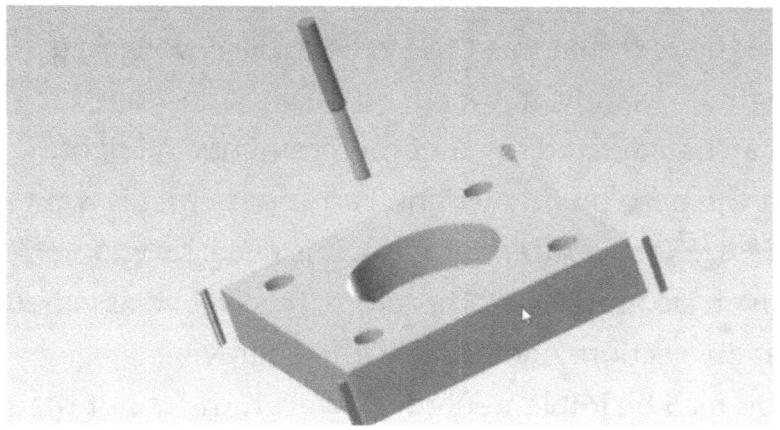

Therefore, we should proceed to the next operation and return to the master. Therefore, let us assume that I wish to engrave a design on this surface using machining. Therefore, what can I do in that situation? Arrive. Additionally, navigate to the wireframe. Initially, we should conceal these tool components. Additionally, the selected tool box is displayed solely by selecting the name. Therefore, at this moment. Marlin has been chosen. Therefore, proceed to the wireframe. Additionally, proceed to the plans. This is a positive development from Solid Phase. Additionally, incorporate this phase. It is a to. Therefore, we should develop the instrument component. Please provide me with a stock of. We will now proceed to merge the aircraft from the solid phase and select it as acceptable. Name. My name is W. He is boarding an aircraft. Additionally, G is generated. Therefore, let us assume that I am the one who is writing. Therefore, I will proceed to the wireframe letters and compose the text that must be present. He is obligated to. All right. Presently, it is inadequately proportioned. Height can be adjusted to meet requirements. All right. Reselect the base point. All right. I anticipate that it will be flawless. Exactly. Afterward. No, it is necessary to establish an isolated layer. Therefore, this will be assigned to the utility section. Additionally, the engraving operation may be implemented. However, this is not the case. The control is user-friendly. We will execute it. Using which. Initially, we will address the issue of control. Therefore, we will be responsible for providing support for the C plane and windows in this area. Within this window. All right. Please provide it. False. Subsequently, which one is this? Simple. Additionally, I will reach the age of 85. Right here. It is feasible for you to execute. Within this location. And proceed to the fifth frame and execute the action within this window. This is the one. And now it is requesting the start points, or this is the start point, which is that we have received our change. Additionally, we will implement an instrument to accomplish this. However, let us apply a filter. Spot drilling is feasible. Or something similar to metal and flat. Therefore, we should select the plate and then compare it to the other. I will perform 0.5. Afterward. All right. We have not explicitly instructed you to perform this action or have we simply generated it in the metal cutting diameter of 0.5? Subsequently, the overall length is five without a trimming band. Afterward, the shoulder extends to the arm, and the diameter of the shoulder is one. Point five will appear more appealing. This one may be acceptable, as it has the same diameter. It is consistent with either five or three. Alternatively, for. This is clarified in this section. Initially, this is the corner. It is, with no pointed corners. This is a chamfer. All right. This is and this is, one that is angled. I am referring to the chamfer that is one line in distance, as well as the radius and full radius. This is the case, and the radius is visible in the lower corner. It is, it is. This is where we can achieve the mean neck diameter. Subsequently, we shall proceed in the same manner. Subsequently, conclude. All right. Our utilization of this instrument will be implemented. All right. Eliminate the parameter. Allow it to remain in this state. Cut in depth. Multiple cuts with greater depth. Regularly, this is more expeditious. You are aware that the profundity of the break through and link parameters will be, say, one less. All right. Therefore, it is now. Subsequently, proceed two sections. Additionally, we possess this. All right. I will now verify whether I have selected this operation in the simulation. This will be the sole operation that is demonstrated. Conformity. The instrument is inadequately proportioned. All right. Nothing is. I was unable to move. Okay, let us verify our operations. See that? We are. We will merely draw a line. Subsequent operation. Subsequent operation. The compartment. We proceed to the previous section. Is scheduled to commence. Additionally, this is intended for the subsequent operation. Next. Okay, so the subsequent operation. It is not plainly written when we focus in, as some of the chain is not machine-made. All right. This is due to the fact that the instrument is significantly larger than the letters. Therefore, we should simply enhance the letter's magnitude. I will write in order to achieve this. Initially, we should deactivate the device. We indicate. In close proximity. All right. Also, wireframe. Let's remove this plane before this one. Delete this. Okay, so. I am in agreement with its deletion. Let us now resume our writing. Correct the situation. Also in this. Okay, so it is. Opposite. The same. Initially, it is necessary to construct an aircraft. Which are manifested in this instance. Therefore, this is the case. I will immediately activate this.

For it at a later time. Additionally. All right. Please select again. And simply reduce it to seven and. Complete the task. All right. Therefore, navigate to the toolbar and observe. This operation is. It is imperative that we do so. We are required to turn right. And I will click here and in the current region. All right. I will now select all operations and perform the filthy operation by clicking on this link. Therefore, the unclean operation is selected and regenerated. We can regenerate all selected operations in this instance, and this option is the most effective type of regeneration. Depends. All right. Presently, we. Exactly. All right. Select "geometry." All right. All right. I understand. That is impressive. Just one small point machine and select. In order to ensure that it is compatible, proceed to the subsequent operation. Therefore, to accommodate the subsequent step, it is acceptable. spot available. Pocket. Additionally, we possess one. To increase the pace. In the same vein, you are required to. I am aware. Viewing is feasible. Everything is impeccably organized. The entire letter "e" is omitted in this instance due to the fact that the resulting area will be exceedingly minuscule. That is the case. All right. Exactly. All right. And now, suppose you wish to gently chamfer the fillets forward on this and these pointed corners. "And if you are able, you can turn and select 3D in the computer." Indeed. And then this one. Next, navigate to the utility. I suggest that we employ this engraving instrument for this purpose. In reality, parameters. I will move the M to the required field. The option is available here. It is a two-dimensional chamfer. If I select this, we will utilize a tool to verify that the texture is satisfactory. Under what conditions is 2D the first, followed by the linear path? In 2D, the centroid is located at the bottom. The chamfer cuts in a manner that some of the portion will be centered at the side, while some of the portion will be corrected by the to. All right. Therefore, the bottom offset is equivalent to the chamfer, which is 0.5. The bottom offset of the two is less. We would like to employ a value of 0.1 or 0.2. I am unaware of the precise dimensions. If an error occurs, you will typically link the profundity and depth to the point in a positive manner. Consequently, the profundity will be devoid of any substance. All right. Let us establish the summit of the stone. That is the summit of star zero zero. Additionally, the profundity will be expressed as both plus and minus. All right. Select this item and retrieve the utility. okay. from a similar source. I harbored an intense affection for you. And now. Let us investigate. All right. Observe that it is cutting this section at the specified depth, and the edges will be cut through. Certainly, let us verify its arrival. Simulate and navigate to the machine to simulate. All right. It is the sole instance in which the entire cargo was utilized in the most recent operation. So, simulate. Proceed to the subsequent operation. The following, the following, and the following. Next. The subsequent action will originate from the. Yes. Perfect. The depth, depth running in, and breadth are defined. Small point seven. So, this is where we can. All right. This and this. It can be defined as its position in the center. Consequently, the parameter and disable in this line iteration are acceptable. Afterward, I will verify it. Yes. It is evident. Yes. Tool commencing with this one. We should verify it through simulation. Therefore, proceed to simulate and and. At the conclusion. I believe we have chosen a solitary operation. What will transpire during the simulation? The. Subsequent in sequence. Additionally, the preceding operation. Additionally, the preceding instrument initiated operations from this juncture. And now, let us observe. We should commence this process from the corner. All right. We will employ the fit procedure to modify one item. Suppose we activate lead in and lead out. Enter the exited midpoint and close the control in this instance. Therefore, this option was causing me to encounter the midpoint. No problem.

www.ingramcontent.com/pod-product-compliance
Lightning Source LLC
Chambersburg PA
CBHW071018240526
45469CB00006BD/1970